TEUBNER-TEXTE zur Informatik   Band 3

U. Glässer

A Distributed Implementation of Flat Concurrent Prolog on Message-Passing Multiprocessor Systems

TEUBNER-TEXTE zur Informatik

Herausgegeben von
Prof. Dr. Johannes Buchmann, Saarbrücken
Prof. Dr. Udo Lipeck, Hannover
Prof. Dr. Franz J. Rammig, Paderborn
Prof. Dr. Gerd Wechsung, Jena

Als relativ junge Wissenschaft lebt die Informatik ganz wesentlich von aktuellen Beiträgen. Viele Ideen und Konzepte werden in Originalarbeiten, Vorlesungsskripten und Konferenzberichten behandelt und sind damit nur einem eingeschränkten Leserkreis zugänglich. Lehrbücher stehen zwar zur Verfügung, können aber wegen der schnellen Entwicklung der Wissenschaft oft nicht den neuesten Stand wiedergeben.

Die Reihe „TEUBNER-TEXTE zur Informatik" soll ein Forum für Einzel- und Sammelbeiträge zu aktuellen Themen aus dem gesamten Bereich der Informatik sein. Gedacht ist dabei insbesondere an herausragende Dissertationen und Habilitationsschriften, spezielle Vorlesungsskripten sowie wissenschaftlich aufbereitete Abschlußberichte bedeutender Forschungsprojekte. Auf eine verständliche Darstellung der theoretischen Fundierung und der Perspektiven für Anwendungen wird besonderer Wert gelegt. Das Programm der Reihe reicht von klassischen Themen aus neuen Blickwinkeln bis hin zur Beschreibung neuartiger, noch nicht etablierter Verfahrensansätze. Dabei werden bewußt eine gewisse Vorläufigkeit und Unvollständigkeit der Stoffauswahl und Darstellung in Kauf genommen, weil so die Lebendigkeit und Originalität von Vorlesungen und Forschungsseminaren beibehalten und weitergehende Studien angeregt und erleichtert werden können.

TEUBNER-TEXTE erscheinen in deutscher oder englischer Sprache.

A Distributed Implementation of Flat Concurrent Prolog on Message-Passing Multiprocessor Systems

Von Uwe Glässer

Universität-Gesamthochschule-Paderborn

B. G. Teubner Verlagsgesellschaft
Stuttgart · Leipzig 1993

Dr. rer. nat. Uwe Glässer

Born in 1959 at Dormagen near Cologne. Studied Computer Science at Paderborn University from 1981 to 1987. Since 1987 he is with the Dept. of Mathematics and Computer Science of Paderborn University, where he received the Dr. rer. nat. degree in 1992. Fields of interest: Parallel and distributed computing, decentralized system architectures.

Dissertation an der Universität-Gesamthochschule-Paderborn im Fachbereich Mathematik/Informatik

Die Deutsche Bibliothek – CIP-Einheitsaufnahme

Glässer, Uwe:
A distributed implementation of flat concurrent prolog on
message passing multiprocessor systems/von Uwe Glässer. –
Stuttgart; Leipzig: Teubner, 1993
   (Teubner-Texte zur Informatik; Bd. 3)
   Zugl.: Paderborn, Univ., Diss., 1992
   ISBN 978-3-322-97612-3   ISBN 978-3-322-97611-6 (eBook)
   DOI 10.1007/978-3-322-97611-6
NE: GT

Umschlaggestaltung: E. Kretschmer, Leipzig

Contents

Acknowledgements

The work reported here has been carried out during my affiliation with the Department of Mathematics & Computer Science at Paderborn University from June 1987 to May 1992. Part of this work has been funded by the Heinz Nixdorf Institute and the North Rhine-Westphalia (NRW) Joint Research Project on Parallel Computing.

I would like to express my gratitude to the following people, who have contributed to this thesis:

To Prof. Dr. Franz Josef Rammig, who initiated this work, for his advice, support, and helpful criticisms throughout the time.

To Prof. Dr. Hans Kleine Büning for his useful comments from which the thesis has benefited.

To Prof. Dr. Uwe Kastens for stimulating discussions that have led to a much deeper understanding of the formal background on concurrent logic programming.

To all members of the "Parallel Logic Programming Group", I have been working with, for participating in the implementation work.

Finally, I should like to thank my wife Andrea for her patience and encouragement in stages when this was most valuable.

Chapter 1

Introduction and Overview

1.1 General Introduction

Concurrent logic programming has become a favorite approach in the eager attempt to combine parallel computer architectures with high-level programming languages. In recent years, a variety of concurrent logic programming languages together with numerous novel programming techniques have been developed and investigated. Main research activities thereby concentrate on applications concerned with system programming, as these languages are especially suitable for expressing concurrent behaviour and implementing parallel algorithms.

Compared to conventional logic programming, the computational model of concurrent logic programming is more closely related to functional programming. Nevertheless, concurrent logic programming languages provide a number of features inherent to logic programming but not directly available in functional programming languages. First of all, this refers to the unique properties of logical variables and the way these can be utilized for specifying communication and synchronization.

With the aim to investigate the possibilities of embedding concurrent logic programming languages into multi-Transputer hardware environments, the language Flat Concurrent Prolog (FCP) was chosen as an implementation candidate. Due to its expressiveness, FCP is representative for a particularly important subclass of concurrent logic programming languages in the sense that it offers about all of the high-level programming techniques made available by any of the other languages of the same class.

The suggested approach for embedding the parallel execution model of FCP onto a non-shared-memory multiprocessor system relies on the design of a *virtual parallel machine* executing compiled FCP code. Substantial design issues with regard to the underlying parallel FCP machine architecture address the realization of the distributed reduction algorithm, dynamic load balancing, as well as various techniques related to distributed control.

Beside exploitation of maximum parallelism, a main objective was to gain maximum scalability. Most of the investigated concepts are embodied in a prototype implementation running on a fully reconfigurable Transputer system with up to 320 processing nodes.

Parts of the material presented here, are also included in a number of publications: [Glaesser90a], [Glaesser90b], [Glaesser90c], [Glaesser91a], [Glaesser91b], [Glaesser92].

1.2 Significance of the Work

The work presented here contributes to the two scientific fields of *concurrent logic programming* and *massively parallel computer architectures*. The practical value of the proposed design and implementation concepts has been successfully demonstrated by the running prototype implementation of the parallel FCP machine. The noticeable results of the work may be summerized as follows:

- ☐ The obtained prototype provides the first real implementation of an FCP-like language on a synchronous message-passing multiprocessor architecture. Special efforts have been devoted to efficiently integrate the asynchronous communication model of the application language, which rests on communication through shared logical variables, into the synchronous message-passing communication model of Transputer systems.

- ☐ Running on a system with up to 320 processing nodes, the above implementation is unique. It should be noticed that the dimension of the processor network by which a distributed computation is carried out is meaningful for system evaluation, system debugging, and system optimization. As such a system behaves non-deterministically, generating configurations of computation states that depend on the accidental distribution of data, the effect of design or implementation errors as well as advantages and drawbacks of certain design or implementation concepts sometimes become visible with large-scale networks, only.

- ☐ An efficient scheme for distributed data representation – especially for representing globally shared logical variables within an environment where competing binding operations are allowed – together with the necessary protocols controlling its dynamic evaluation is introduced.

- ☐ A new concept of distributed control for treating respectively avoiding livelocks and deadlocks is discussed. The value of diverse known strategies for alternative realizations of dynamic work load balancing on large networks has been investigated.

☐ The proposed realization of the functional architecture for the parallel FCP
  machine follows a hierarchical design methodology offering a much better understanding of the relatively complex distributed algorithm which is necessary
  to embed the language's parallel execution model on Transputer networks. A
  number of hierarchically structured layers of the virtual machine according to
  different levels of abstraction as well as the corresponding interfaces between
  these layers are identified. Furthermore, the design as a whole provides a maximum of scalability and flexibility with regard to the size and the topology of the
  underlying processor network.

1.3 Overall Organization

The thesis consists of seven chapters corresponding to the following organization:

The remainder of Chapter 1 contains a few elementary definitions referring to frequently used notations of logic programming. Readers which are
familiar with the subject may skip this part without loss of continuity.

Chapter 2 introduces the computational model of concurrent logic programming and the concurrent logic programming language FCP in terms of its
operational semantics.

Some basic concepts for implementing FCP on uniprocessor machines, as
they are embodied in the execution model of a sequential FCP machine,
are briefly discussed in Chapter 3.

A detailed description of the parallel FCP machine is presented in Chapter
4 and Chapter 5. With respect to an abstract system architecture for a
message-passing multiprocessor system, Chapter 4 introduces the fundamental parallelization concepts. Chapter 5 thereupon specifies a concrete
functional architecture for implementing the parallel FCP machine on large-
scale Transputer networks.

Chapter 6 provides performance measures obtained from a prototype implementation running on a fully reconfigurable multi-Transputer system.

The final chapter contains the conclusions.

1.4   Syntax of Logic Programs

Definitions

The notation used for representing logic programs throughout this work essentially follows the usual conventions of *PROLOG* ([Clocksin84],[Kleine Buening86]) which go back to the Edinburgh syntax [Bowen81]. Nevertheless, having some basic definitions at least for the most frequently referred expressions might be helpful:

Logical Variable.   A logical variable is either represented by a single capital letter, a single underline '_', or it is represented by an identifier with an initial capital letter or an initial underline (e.g. *X, Xs, _12*).

Term. Terms are the universal data structure for representing individual data objects. A term is either a variable or an expression of the form $f(T_1, T_2, ..., T_n)$, where f is a *function symbol* of arity n and the T_i are terms, or it is a list.

A term is *atomic* if it is a variable or a constant, i.e. a function symbol of arity $n = 0$, or the empty list [ ]. A term is *compound*, if it is either of the form $f(T_1, T_2, ..., T_n)$, or of the form $[T_1, T_2, ..., T_n]$, or of the form $[[T_1, T_2, ..., T_n | Xs]$, where $n > 0$, $T_1, T_2, ..., T_n$ are terms, f is a function symbol of arity n, and the *tail variable* X represents an unknown tail list. Terms that do not contain any variables are denoted as *ground terms*.

A list structure of the form $[a| [b| [c| [ ]]]]$ is usually rewritten as $[a, b, c]$.

Examples.

Atomic terms:	$X, \quad a, \quad 1.5, \quad [\,], \quad Variable$
Compound terms:	$f(X, Y), \quad g(a, Z, 2, h([\,])),$
	$[1, 2, 3| [\,]], \quad [X|Xs]$

Atom. An atom is an atomic formula of the form

$$p(T_1, T_2, ..., T_k), \quad k \geq 0,$$

where p is a predicate symbol of arity k and $T_1, T_2, ..., T_k$ are terms in the argument of p. The predicate is identified by the expression p/k. A conjunction of atoms $A_1 \wedge A_2 \wedge ... \wedge A_n$ is represented as $A_1, A_2, ..., A_n$.

Program Clause. A single program clause has the general structure 'Head $\leftarrow$ Body.' The *Head* consists of a single atom A, while the *Body* consists of a possibly empty conjunction of atoms $B_1, B_2, ..., B_n$ $(n \geq 0)$.

An empty conjunction of atoms is represented by the constant *true*. A clause C of the form $C = 'A \leftarrow true'$ is called a *unit clause*. Unit clauses represent *facts*.

Let $X_1, X_2, ..., X_k$ denote the variables occurring in the clause atoms $A, B_1, B_2, ..., B_n$. Then a clause C is a universally quantified formula of the form

$$(\forall X_1, X_2, ..., X_k)\ (A \leftarrow B_1, B_2, ..., B_n), \quad (n \geq 0).$$

A clause of this type is called a *Horn clause*. Since all clauses are quantified in the same way, the quantification is assumed to be defined implicitly; hence, it is usually dropped in the notation.

Logic Program. A logic program $\mathcal{P}$ is a finite set of universally quantified Horn clauses $\mathcal{P} = \{C_1, C_2, ..., C_s\}$, $s > 0$.

Program Procedure. With respect to a logic program $\mathcal{P} = \{C_1, ..., C_s\}$, a *program procedure* $C^{p/k}$ is the particular subset of clauses in $\mathcal{P}$ collectively defining the predicate p/k, $C^{p/k} = \{C_1^{p/k}, C_2^{p/k}, ..., C_r^{p/k}\} \subseteq \mathcal{P}$, $(r \leq s)$. More precisely, within the program $\mathcal{P}$, $C^{p/k}$ identifies all clauses with the common head predicate symbol p which is of arity k.

Computation Goal. A computation goal is a conjunction of atoms $A_1, A_2, ..., A_n$ $(n \geq 0)$. The goal is *atomic* if $n = 1$. The goal is *empty* if $n = 0$. The individual A_i are also denoted as *goal atoms*.

1.5   Data Manipulation by Unification

The basic data manipulation operation when executing a logic program is unification. Compared to more conventional data manipulation mechanisms, e.g. as applied by imperative programming languages, the assignment of values to variables by means of unification shows rather different properties.

According to the *single-assignment property* of logical variables, a variable can be assigned a value at most once. In fact, this is not as restrictive as it might seem, since the value assigned to a variable itself can be another variable as well as a compound term containing other variables.

The unification of two terms T and T', informally, can be described as an operation trying to replace the variables in T and T' by appropriate terms in such a way as T and T' become identical. A successful unification computes a substitution set θ specifying the affected variables together with the corresponding terms they have to be replaced with. The unification fails if such a substitution θ does not exist for the terms T and T'.

Definition. Let T be a nonground term and $X_1, X_2, ..., X_n$ be the variables occurring in T. A *substitution* θ for the term T is a finite set of the form $\theta = \{(X_{i_1}, T_1), (X_{i_2}, T_2), ..., (X_{i_k}, T_k)\}$, where $\{X_{i_1}, X_{i_2}, ..., X_{i_k}\} \subseteq \{X_1, X_2, ..., X_n\}$ and the T_j are terms. Each pair (X_{i_j}, T_j) defines a replacement of a variable X_{i_j} by a term T_j satisfying the conditions defined below

$$(\forall g)(\forall h)\; g, h \in \{1, ..., k\} :$$

$$(1) \quad g \neq h \;\Rightarrow\; X_{i_g} \neq X_{i_h} \quad and$$
$$(2) \quad X_{i_g} \text{ does not occur in } T_h$$

The expression $T\theta$ refers to the term which is obtained as the result of applying substitution θ on term T. $T\theta$ denotes an *instance* of T.

Definition. The unification of two terms T and T' is defined by the relation *unify* in the following way

$$unify(T, T') = \begin{cases} \Theta = \{\theta \mid T\theta = T'\theta\}, & \text{if } \Theta \neq \emptyset \\ fail, & \text{otherwise.} \end{cases}$$

If Θ is non-empty, then the terms T and T' are *unifiable*. The elements of Θ are called the *unifiers* of T and T'.

Definition. If two terms T and T' are unifiable with the unifier set Θ, then there is at least one *most general unifier (mgu)* θ, $\theta \in \Theta$, with the following property:

$(\forall \theta')\ \theta' \in \Theta - \{\theta\} :$

    (1)   For the term $T\theta$, there is a substitution σ, such that $(T\theta)\sigma = T\theta'$, i.e. $T\theta'$ is an instance of $T\theta$; *or*

    (2) θ and θ' are identical up to a renaming of variables.

Condition (2), in particular, means that θ' is a mgu of T and T' as well.

Definition. When considering the unification of two terms T and T', a unification with a most general unifier is requested. In general, if T and T' have the unifier set Θ containing a subset Θ' of most general unifiers then it does not concern which one out of several possibly existing most general unifiers is computed.

For that reason, the result of a successful unification of T and T' yielding some *mgu* θ is simply referred as $mgu(T, T') = \theta$ instead of $\theta \in mgu(T, T')$. In the complementary case, if T and T' do not unify, this is indicated by $mgu(T, T') = fail$.

Example 1. $\quad T = f(X, Y), \quad T' = f(a, Z)$

$$mgu(T, T') : \quad \theta_1 = \{X \to a, Y \to Z\},$$
$$\theta_2 = \{X \to a, Z \to Y\}$$

Example 2. $\quad T = f([X|Xs], g(Ys), Ys),$

$$T' = f([1, 2|Ys], Z, Zs)$$

$$mgu(T, T') : \quad \theta_1 = \{X \to 1, Xs \to [2|Zs],$$
$$Ys \to Zs, Z \to g(Zs)\},$$
$$\theta_2 = \{X \to 1, Xs \to [2|Ys],$$
$$Zs \to Ys, Z \to g(Ys)\}$$

Chapter 2

Concurrent Logic Programming

2.1 Process Interpretation of Logic Programs

Process interpretation of logic programs establishes the abstract computational model of concurrent logic programming in which the active objects of a computation are conceived as *concurrent processes*. Process interpretation, generally, is in contrast to procedural interpretation of logic programs [Ueda89]. The latter one provides the abstract computational model of conventional (sequential) Prolog, as introduced by Kowalski in ([Kowalski79a],[Kowalski79b]).

2.1.1 The Process Model

Process interpretation of logic programs means that the conjunctive goal of a computation is regarded as an asynchronous *process network*. The concurrent processes communicate and synchronize via shared logical variables according to an asynchronous communication model. Shared logical variables thus form interprocess communication channels and by that means they reflect the network's interconnection structure. Based on this concept various kinds of communication protocols can be implemented. A number of illustrative paradigms e.g. is presented in ([Shapiro86],[Shapiro89]).

An asynchronous communication model implies the following process behaviour. A process reads from input variables and writes onto output variables in such a way as process execution is blocked until all the values required from process input variables become available. When being executed, a process may produce results by instantiating output variables. Processes consuming values from these output variables thereby become activated.

In the computational model of concurrent logic programming unification realizes both, the basic *data manipulation operation* as well as the elementary *communication primitive*. Moreover, it allows to describe and to control communication implicitly by means of ordinary language constructs rather than explicitly by additional communication control constructs.

Process interpretation of logic programs was introduced by van Emden and de Lucena Filho [vanEmden82]. However, they used a model which is different from the actual process model of concurrent logic programming. In the original model, a process has been considered as a sequential unit of computation executing subgoals from a local stack. Communication channels connect individual subgoals belonging to different processes of the same computation. A process therefore forms an abstract functional block that may run in parallel with other processes but has its own sequential control structure.

Compared to the model described above, the actual process model of concurrent logic programming uses more light-weighted processes and as a consequence thereof, it provides more fine grained parallelism. A process always corresponds to a single computation goal, which is represented by a goal atom of the form

$$p(A_1, A_2, ..., A_k) \quad (k \geq 0).$$

The goal atom's predicate symbol p/k identifies the process *program state*, where k denotes the arity of p. The list of terms $A_1, A_2, ..., A_k$ in the argument of p/k is interpreted as a collection of process registers reflecting the current process *data state*.

2.1.2 Computational Behaviour

A process can perform a single operation called *process reduction*. When to reduce a process and how to reduce it, depends on the actual process argument values, i.e. the process data state.

Regarding a logic program $\mathcal{P}$, the operational behaviour of a process $p(A_1, ..., A_k)$ is determined by the finite subset $C^{p/k}$ of program clauses, $C^{p/k} = \{C_1{}^{p/k}, ..., C_l{}^{p/k}\} \subseteq \mathcal{P}$ $(l > 0)$, collectively defining the predicate p/k. The clauses of $C^{p/k}$ are considered as a *program procedure*. Each clause $C_i{}^{p/k} \in C^{p/k}$ $(i \leq l)$ represents a rewrite rule for goal atoms of type p/k. It has the general from of a *guarded Horn clause*, where A, the G_i, and the B_j respectively represent atoms (as explained below):

$$\underbrace{A}_{Head} \leftarrow \underbrace{G_1, G_2, ..., G_m}_{Guard} \mid \underbrace{B_1, B_2, ..., B_n}_{Body} \quad (m, n \geq 0)$$

The *guard part* $G_1, G_2, ..., G_m$ is read as a conjunction of predicates controlling clause selection. Guard predicates state conditions referring to process arguments. In order to reduce a process A' using the program clause $A \leftarrow G_1, ..., G_m \mid B_1, ..., B_n,$

the goal atom A' must unify with the clause's head atom A and, in addition, evaluation of the guard must succeed; that is, all the guard predicates $G_1, ..., G_m$ must be fulfilled simultaneously. The usage of guards in guarded Horn clauses with respect to the resulting impact on program control is closely related to the effect of a guard in the *alternative construct* of Dijkstra's guarded command [Dijkstra75].

The *body part* $B_1, B_2, ..., B_n$ is read as a collection of atoms specifying a set of *concurrent subprocesses*. As a result of a successful reduction step, these subprocesses spawn a local subnetwork replacing the reduced process within the global process network. Due to unification of respective variables in the environment of the process and the newly created subprocesses, the data state of the resolvent is updated accordingly.

A process reduction step is *enabled* if the related program procedure contains one or more *enabling* clauses, i.e. clauses that apply to reduce the process. In order to find an enabling clause, all the clauses of the program procedure may be tried in parallel. When there are several clauses applicable at a time, the commit operator "|" acts as a control primitive ensuring that clause selection is carried out in a *mutual exclusive* manner.

Upon successful head unification and guard evaluation, a clause is definitively chosen for reduction as soon as the evaluation procedure reaches the commit operator. At the same moment, all alternative choices concurrently being regarded are discarded. The potential for parallelism offered by the clause selection procedure might be considered as some form of restricted OR-parallelism.

According to their dynamic behaviour, we can identify two basic kinds of processes: *iterative processes* and *general processes*. Thereby, the behaviour of a process when being reduced primarily depends on the structure of the reducing clause. Referring to the implied process behaviour, all clauses can be classified into three different categories, namely:

$$
\begin{array}{lll}
\textit{general clauses:} & A \leftarrow G_1, G_2, ..., G_m \mid B_1, B_2, ..., B_n & (m \geq 0,\ n \geq 2) \\[4pt]
\textit{iterative clauses:} & A \leftarrow G_1, G_2, ..., G_m \mid B_1 & (m \geq 0,\ n = 1) \\[4pt]
\textit{unit clauses:} & A \leftarrow G_1, G_2, ..., G_m \mid true & (m \geq 0,\ n = 0)
\end{array}
$$

A general clause specifies a *process fork* into a network of concurrent subprocesses. An iterative clause specifies a *state transition* modifying process argument values. A reduction by means of a unit clause results in a *process termination*. The term *true* represents the empty process network.

The Role of Nondeterminism. The computational behaviour resulting from the execution model described above is called *guarded-command indeterminacy* [Dijkstra75] or *don't-care nondeterminism* [Kowalski79b] in contrast to *don't-know nondeterminism* as applied in the computational model of Prolog. This terminology refers to two different ways of interpretating nondeterminism in computations performed by logic programs.

Depending on the kind of nondeterminism utilized for program execution, the way computations may interact with the "outside world" is substantially different. Don't-care nondeterminism offers the possibility to produce *partial results* (partial answer substitutions) without knowing whether a computation will succeed or fail. Don't-know nondeterminism, however, means that a result is produced only in case of succeeding computations and for that reason it cannot be produced before the mode of termination is known. This feature has a strong impact on the kind of systems that can be specified by the various logic programming languages. The ability of concurrent logic programming languages to produce partial results, in particular, is essential for the specification of *reactive behaviour* in concurrent systems ([Harel85],[Shapiro89],[Ueda89]).

2.1.3 General Language Classification

With respect to the complexity of guard predicates, concurrent logic programming offers two distinct approaches ([Shapiro89], [Takeuchi87]). A more general approach permits the usage of arbitrary complex guard predicates. In particular, it allows guard predicates themselves to be defined by means of program procedures of the application program.

As a consequence, arbitrary complex subprocesses may be spawned during guard evaluation and the reduction of these subprocesses in turn may require the evaluation of similarly complex guards. The execution of a program, this way, can result in an *unbounded hierarchy* of guard process calls, which necessitates the realization of rather complex control mechanisms. For example, such a complex control mechanism would be required in order to be able to perform distributed commit operations. Languages following this approach, e.g. are *Concurrent Prolog* [Shapiro83], *PARLOG* ([Clark84],[Gregory87]), and *Guarded Horn Clauses (GHC)* [Ueda86].

Alternatively, so-called *flat languages* restrict the usage of guard predicates on a *predefined* set of primitive and built-in *test operations*. The advantage of this approach, in contrast to nonflat languages, results from a significant decrease in program control complexity. Guard test predicates can always be evaluated, immediately, i.e. without performing complex subcomputations. In the execution of a flat language, the goal therefore corresponds to a flat collection of processes.

For almost all concurrent logic programming languages flat language subsets have been defined and studied. These are considered to be more suitable for efficient implementations due to their simplicity, while the loss of expressiveness when restricting on the flat subset of a concurrent logic language seems to be relatively small. Examples of flat languages are *Flat Concurrent Prolog (FCP)* [Mierowsky85], *Flat GHC*, *Flat Parlog* [Foster87], and *Strand* [Foster90]. A detailed overview on flat languages is presented in [Shapiro89].

2.2 Data-Flow Synchronization Techniques

The following discussion on data-flow synchronization techniques of concurrent logic programming languages focuses on flat languages in general and FCP in particular. However, most of the features being addressed are relevant for non-flat languages, as well.

2.2.1 Process Synchronization

The basic synchronization concept of concurrent logic programming is to delay reduction operations depending on the instantiation state of variables in the arguments of the processes to be reduced. This is realized by means of a *process suspension mechanism.*

An attempt to reduce a process causes a *suspension* of that process if it is not possible to decide, whether the process can be reduced or not. That means, as the argument variables of the process have not been sufficiently instantiated so far, it is neither possible to determine a reducing clause nor to recognize that none of the potential program clauses is applicable at all. A suspension therefore indicates that certain input data required for the process to be reduced is not yet available but can probably be obtained as the result of reducing some other processes.

When the corresponding variables in the environment of the process become further instantiated, a repeated reduction attempt may have a definite result; the process is either reduced successfully or, otherwise, a *reduction failure* occurs. In the later case neither of the clauses would enable a reduction regardless of whatever the process variables are instantiated to.

In case of a successful reduction operation, data is communicated between the process being reduced and the subprocesses replacing it, as respective variables are unified. Process variables can either be instantiated to nonvariable terms or they can be bound with one another. However, as the outcome of a reduction attempt is not known in advance, it must always be guaranteed that any variable bindings that have been computed prior to commitment to a clause can be *undone* properly. If an attempt to reduce a process with a chosen program clause fails, the process variables must remain unchanged.

2.2.2 Atomicity of Unification

A main difference in the computational models of the various concurrent logic programming languages results from the distinct granularities of their atomic data manipulation operations. Though almost all of these languages perform data manipulation by means of unification, the complexity of the unification operations that are carried out as an indivisible computation step is quite different.

The ability to unify arbitrary complex terms within a single unification operation is referred to as *atomic unification*. Any weaker form of unification is referred to as *non-atomic unification*. FCP for instance applies atomic unification, whereas Flat GHC only guarantees that the instantiation of a variable to a value is carried out as an atomic operation. In fact, this means that Flat GHC performs a compound unification of the form $f(S_1, S_2, ..., S_k) = f(T_1, T_2, ..., T_k)$ in such a way as if the unification operation would be broken up into a number of related subunifications which are represented by the goal $S_1 = T_1, S_2 = T_2, ..., S_k = T_k$ [Shapiro89].

The granularity of a language's atomic operations has a substantial impact on the synchronization mechanisms required by its parallel implementation. The finer the granularity, the more simple the synchronization mechanisms to be realized.

2.2.3 Specification of Synchronization

Apart from using atomic operations of different granularities, another important criterion by which concurrent logic programming languages are characterized, is the way of how process synchronization is specified within a program clause. The unification of a process with the head of a clause can be carried out in several different ways, which have a considerable effect on the achieved synchronization behaviour [Takeuchi87].

The most restrictive form of head unification is called *input matching* or just *matching*. Input matching realizes some kind of one-way unification, where the direction of assigning bindings to variables is predefined. Unification prior to commitment to a clause must not affect variables in the environment of the process; otherwise, the process becomes suspended. Any binding of a process variable with another variable or a nonvariable term therefore has to be specified within the body part of a clause. Actually this means that the reduction of a process A' by means of a reducing clause $'A \leftarrow G \,|\, B'$ is always delayed until A' has become an instance of A. Input matching is applied in GHC and Flat GHC.

An alternative solution instead of using input matching was chosen for PARLOG and the languages derived thereof. Static *mode declarations*, which are part of the declaration of program procedures, restrict the access mode of each process argument to be either *input* or *output*. When performing head unification, process variables occurring at the position of an input argument are protected against write operations, similar to input matching.

The most general form of head unification is *read-only test unification* as applied by Concurrent Prolog and FCP ([Mierowsky85],[Shapiro86]). Read-only unification is an extension of general unification based on the concept of read-only variables. Using the *read-only operator '?'* as a data-flow synchronization primitive, various occurrences of the same logical variable can be supplied with different access modes. From an ordinary write-enabled variable $'X'$ we obtain the corresponding read-only variable $'X?'$.

In contrast to write-enabled variables, there is no write access to read-only variables, except by instantiating their write-enabled counterparts. During head unification, any attempt to unify a process read-only variable with a data object other than a write-enabled variable results in a process suspension. The unification of a read-only variable with a write-enabled variable, however, is permitted since this does not affect the read-only variable itself.

Compared with the other approaches, the concept of read-only unification realizes a dynamic way of specifying synchronization because of the fact that the access mode is attached to the data objects. Synchronization via input matching or PARLOG-like mode declarations is of static nature, as the access mode is specified by the program procedures.

2.2.4 Overall Comparison

The various approaches for realizing communication and synchronization have led to a diversity of associated (flat) concurrent logic programming languages. Each of these is characterized by using some combination of the following concepts: *atomic unification* versus *non-atomic unification*, and *static synchronization* versus *dynamic synchronization* [Taylor89].

In general, there seems to be a trade-off between simplicity at the implementation level versus expressiveness at the language level. The more expressive a concurrent logic programming language is, e.g. by offering features like atomic unification or dynamic synchronization, the more complex is its implementation.

How much expressiveness and elegance with respect to high-level logic programming techniques is really needed for practical applications, is still an open question. Nevertheless, practical experiences in the field of system programming have shown that less expressive languages need to be augmented by additional control primitives. These are required in order to monitor and control the status of computations, e.g. to reflect on the termination and failure of a computation.

The language $KL1$[1], for example, was obtained from Flat GHC by adding a number of *meta-programming* functions used for meta-level control, such as starting or aborting computations, resource management, scheduling, and load distribution [Nakajima92].

A more expressive language like FCP has already the potential ability to provide the necessary control functions without relying on additional control primitives. Though, it is not known if these functions can be realized with the necessary efficiency. Apart from this, almost all of the simpler languages, including KL1, have a simple and natural embedding in FCP, as has been demonstrated in [Shapiro89].

2.3 Operational Semantics of FCP

The operational semantics of a programming language is to be defined by an implementation independent algorithm for executing programs written in this language [vanEmden76]. Following this approach, the operational semantics of FCP is explained in terms of an abstract interpreter for uniprocessor execution of FCP programs ([Mierowsky85],[Shapiro86]).

Basically, the execution of an FCP program is described by a number of process reduction steps corresponding to a computation on a given input goal. However, a process reduction is a complex operation that requires to perform one or more clause try operations in order to determine a reducing clause. For that reason, the clause try operation, which also provides the layer for specifying data-flow synchronization, needs to be discussed in some detail, first.

2.3.1 FCP Data-Flow Synchronization

In order to give a precise specification of the behaviour implied by the data-flow mechanism of FCP, two things have to be defined. The first one is the set of *guard test predicates* that are available in FCP. The second one is the *clause try function* controlling clause selection depending on the results of head unification and guard evaluation. This is a commonly used notion when specifying synchronization mechanisms of concurrent logic programming languages. [Kliger88].

---

[1] *KL1 (Kernel Language Version 1)* represents the core language for the *Parallel Inference Machine (PIM)* being developed as part of the Japanese *Fifth Generation Computer Project (FGCP)* at ICOT. Its distributed operating system *PIMOS (Parallel Inference Machine Operating System)* is implemented in KL1 ([Fuchi86],[Takeda90]).

FCP Guard Test Predicates. Below, there is a complete list of FCP guard test predicates as defined in [Silverman87]. The evaluation of a guard test predicate suspends if it cannot be decided whether the predicate succeeds or fails.

Unification:

$X = Y$	X and Y unify.
$X \neq Y$	X and Y fail to unify.
$X =?= Y$	X and Y are identical (unify without assigning variables).

Arithmetic:

$X =:= Y$	Arithmetic comparison of numeric expressions.
$X > Y$	ditto.
$X >= Y$	ditto.
$X < Y$	ditto.
$X =< Y$	ditto.

Type checking:

integer(X)	X is an integer.
real(X)	X is a real number.
number(X)	X is a number (either integer or real).
string(X)	X is a string.
constant(X)	X is a constant (either a number or a string or nil).
tuple(X)	X is a tuple.
list(X)	X is a list.
compound(X)	X is a compound (either a tuple or a list).

Term comparison:

$X @< Y$	X is less then Y according to the canonical ordering: numbers $@<$ strings $@<$ [ ] $@<$ lists $@<$ tuples. Variables are incomparable, so for two different variables X and Y the tests $X @< Y$, $X \neq Y$, and $X =? = Y$ suspend. Strings are ordered lexically, tuples are ordered according to their arity.

Meta-logical:

unknown(X)	the value of X is unknown.
known(X)	the value of X is known.

Control:

true	succeeds.
otherwise	succeeds when it is detected that all textually previous clauses fail.

The Clause Try Function. Presuming the predefined set of guard test predicates given above, the clause try function *try* determines the synchronization behaviour when attempting to reduce a process A' by means of a selected program clause $A \leftarrow G_1, ..., G_m \mid B_1, ..., B_n$. For the try function to be computed, the results obtained by performing two subordinate operations have to be combined and evaluated. These are the unification of the goal atom representing process A' with the clause head A and the evaluation of the guard part. Variable bindings resulting from the unification thereby have to be applied to the guard test predicates $G_1, ..., G_m$.

The result of a clause try operation is either *succeed* or *fail* or *suspend*. While *succeed* and *fail* are stable results, *suspend* is not stable. Depending on further instantiations of variables in the argument of A' a repeated attempt of a suspended clause try may either succeed, fail or suspend again.

When succeeding, a clause try operation of the form

$$try(A', (A \leftarrow G_1, ..., G_m \mid B_1, ..., B_n))$$

yields a substitution $\theta_?$ as the *read-only mgu* of A' and A. For the read-only extension of conventional unification, as introduced in section 1.5, we define the following concepts ([Shapiro86],[Shapiro89]):

Definition. A substitution θ is *admissible* with respect to a term T containing the read-only variables $X_1?, X_2?, ..., X_n?$ if the resulting term $T\theta$ satisfies the subsequent condition:

$$(\forall i)\, 1 \leq i \leq n : \quad (X_i?)\theta = X_i?$$

Definition. A *read-only extension* $\theta_?$ of an admissible substitution θ when applied to a term T containing the variables $X_1, X_2, ..., X_n$ is a substitution satisfying the following two conditions[2]:

$$(\forall j)\, 1 \leq j \leq n :$$

$$(1)\quad X_j\theta_? \; = \; X_j\theta \;\; and$$
$$(2)\quad (X_j?)\theta_? \; = \; (X_j\theta)?.$$

Definition. For two terms T and T' let $\theta = \{X_1 \rightarrow T_1, ..., X_n \rightarrow T_n\}$ denote the $mgu(T, T')$ if T and T' are unifiable. The *read-only mgu* of T and T', which is denoted by $mgu_?(T, T')$, then is defined as

$$mgu_?(T, T') = \begin{cases} \theta_? = \; \theta \cup \; \{X_1? \rightarrow T_1, ..., X_n? \rightarrow T_n\}, \\ \qquad\quad \text{if } mgu\,(T, T') = \theta \text{ and } \theta \text{ is admissible,} \\ fail, \qquad \text{if } mgu\,(T, T') = fail, \\ suspend, \quad \text{otherwise.} \end{cases}$$

Examples.

(1)$\quad T = f(X, g(X?)), \quad T' = f(a, Y)$

(Since X unifies with a, $X?$ is instantiated to a, too. In addition, Y and $g(a)$ can also be unified.)

$$\Rightarrow mgu_?(T, T') = \{X \rightarrow a, \; X? \rightarrow a, \; Y \rightarrow g(a)\}$$

(2)$\quad T = f(X, g(X?)), \quad T' = f(a, g(b))$

(Since X unifies with a, $X?$ is also instantiated to a; but the unification of the resulting term $g(a)$ with $g(b)$ fails.)

$$\Rightarrow mgu_?(T, T') = fail$$

---

[2]When applying the read-only operator '?' to a ground term, it does not have any effect, i.e. the result of this operation is the ground term itself.

$$(3) \quad T = f(X, g(X?)), \quad T' = f(Y, g(a))$$

(It cannot be decided whether the terms $g(X?)$ and $g(a)$ unify, since X is not instantiated to a nonvariable term.)

$$\Rightarrow mgu_?(T, T') = suspend$$

Definition. By means of the function $mgu_?$ and an additional function *test* evaluating the guard to be either *true, false* or *suspend*, the clause try function *try* is defined below:

$$try(A', (A \leftarrow G_1, ..., G_m \mid B_1, ..., B_n))$$

$$= \begin{cases} \theta_?, & \text{if } mgu_?(A, A') = \theta_? \ \wedge \ test\left((G_1\theta_?, ..., G_m\theta_?)\right) = \ true, \\[2ex] fail, & \text{if } mgu_?(A, A') = \theta_? \ \wedge \ test\left((G_1\theta_?, ..., G_m\theta_?)\right) = \ false \\ & \text{or } mgu_?(A, A') = \ fail, \\[2ex] suspend, & \text{otherwise.} \end{cases}$$

2.3.2   An Abstract FCP Interpreter

This section describes a uniprocessor version of an abstract interpreter for FCP. The interpreter is abstract in the sense that it does not specify what is required in any real implementation of it: the process scheduling policy. Instead of, it is assumed that the interpreter applies some *nondeterministic* process selection procedure. Making nondeterministic choices, it never selects a process for which the reduction attempt will result in a suspension. A process suspension mechanism therefore becomes superfluous.

Moreover, a clause selection policy is also not defined. For simplicity, OR-parallel clause selection can be assumed. OR-parallelism, of course, remains restricted to the evaluation of clause try functions because of the effect of the commit-operator. The interpreter algorithm is given below:

INPUT: An FCP program $\mathcal{P}$, and an initial process A.

OUTPUT: Either $A\theta_?$, or *suspend* , or *fail.*
            ($A\theta_?$ represents an instance of A derived from $\mathcal{P}$.)

INTERPRETER:

- $\mathcal{R} := \{A\}$ /* Initialize the resolvent $\mathcal{R}$. */

- While $\mathcal{R} \neq \emptyset$

  - Choose some non-suspending process A' from $\mathcal{R}$
    (If such a process does not exist, then *Exit.*)

  - Choose a clause $A" \leftarrow G_1, \ldots, G_m \mid B_1, \ldots, B_n$ from $\mathcal{P}$,
    such that

    $$(\exists\,\theta_?): \; try(\,A', (A" \leftarrow G_1, \ldots, G_m \mid B_1, \ldots, B_n)\,) = \theta_?$$

    (If such a clause does not exist, then output(*fail*); *Halt.*)

  - $\mathcal{R} := \mathcal{R} - \{\,A'\,\}$

  - $\mathcal{R} := (\,\mathcal{R} \cup \{B_1, \ldots, B_n\}\,)\theta_?$

- if $\mathcal{R} = \emptyset$

    then output($A\theta_?$)
    else output(*suspend*)

- Halt.

2.3.3   Formal Description of Computations

Depending on the granularity of operations being observed, there are a number of distinct layers for describing computations of concurrent logic programs. Basically, these are unification, clause try operations, reduction operations, and the computation as a whole. When referring to control flow and data flow issues in the execution of a program, a particular interesting layer is the one dealing with reduction operations. A suitable means for modeling computations in such a way offer *transition systems* [Pnueli86]. Special transition system for FCP have been proposed by Shapiro [Shapiro89] and Taylor [Taylor89].

A transition system for describing computations of concurrent logic programs consists of a set of *computation states* and a set of *transition rules* specifying mappings from computation states into sets of computation states. The subsequently defined transition system for FCP applies a notation which is similar but not identical to the original definitions mentioned above.

Definition. A *computation state* is a pair $< \mathcal{R} ; \theta >$, where $\mathcal{R}$ is either a sequence of atoms representing the current resolvent or a mode identifier x, $x \in \{succeed, suspend, fail\}$. The substitution θ provides the current set of variable bindings computed so far.

An *initial* computation state is given by a pair of the form $< \mathcal{R}_0 ; \theta_0 >$, where $\mathcal{R}_0$ represents the initial resolvent and θ_0 corresponds to the empty substitution ($\theta_0 = \emptyset$). A pair of the form $< x ; \theta >$ indicates a *terminal state*.

Definition. A *transition rule t* specifies a transition from a state S to a state S' if $S' \in t(S)$. This is written as $S \vdash S'$. A transition is *enabled* on a state S if $t(S) \neq \emptyset$; otherwise, S is a terminal state. Any computation performed by an FCP program $\mathcal{P}$ can be described using the following four transitions rules, where $\theta \circ \theta'$ denotes the composition of θ and θ':

(1) *REDUCE:*

$$S = < A_1, ..., A_{i-1}, A_i, A_{i+1}, ..., A_r ; \theta >$$
$$\vdash_{REDUCE} \ < (A_1, ..., A_{i-1}, B_1, ..., B_n, A_{i+1}, ..., A_r)\theta' ; \theta \circ \theta' >$$

A *REDUCE* transition is enabled on a state S if $\mathcal{P}$ contains a clause $A \leftarrow G_1, ..., G_m \mid B_1, ..., B_n$ such that

$$try(A_i, (A \leftarrow G_1, ..., G_m \mid B_1, ..., B_n)) = \theta'.$$

(2) *SUCCEED:*

$$S = \ < true ; \theta > \vdash_{SUCCEED} \ < succeed ; \theta >$$

(3) *FAIL:*

$$S = \; < \mathcal{R}; \theta > \; \vdash_{FAIL} \; < fail \, ; \theta >$$

The *FAIL* transition is enabled on a state S if $\mathcal{R} \neq true$ and for some A_i in $\mathcal{R}$ and every clause C of $\mathcal{P}$

$$try(A_i, C) = fail \; .$$

(4) *SUSPEND:*

$$S = \; < \mathcal{R}; \theta > \; \vdash_{SUSPEND} \; < suspend \, ; \theta >$$

The *SUSPEND* transition is enabled on a state S if $\mathcal{R} \neq true$ and for all A_i in $\mathcal{R}$ and for every clause C of $\mathcal{P}$

$$try(A_i, C) = suspend \; .$$

Definition. A *computation* $\mathcal{C}$ of an FCP program $\mathcal{P}$ starting from the initial state $< \mathcal{R}_0 \, ; \theta_0 >$ corresponds to a sequence of state transitions $\mathcal{C} = < \mathcal{R}_0 \, ; \theta_0 >, < \mathcal{R}_1 \, ; \theta_1 >, \ldots$. If the computation is finite, it will eventually reach a terminal state $< x \, ; \theta >$. Depending on the value of x, a finite computation is either called a *successful, a failing, or a suspending* computation.

Definition. Equivalence between two finite computations of the same program can be expressed with respect to the *observable behaviour* of these computations ([Shapiro89],[Ueda89]). The observable behaviour of a finite computation $\mathcal{C}$ of the form $\mathcal{C} = < \mathcal{R}_0 \, ; \theta_0 >, < \mathcal{R}_1 \, ; \theta_1 >, \ldots, < \mathcal{R}_{n-1} \, ; \theta_{n-1} >, < x \, ; \theta_n >$, corresponds to a triple $< \mathcal{R}_0 \, ; \theta \, ; x >$, $x \in \{succeed, \; suspend, \; fail\}$. The *answer substitution* θ is obtained by restricting θ_{n-1} to the variables in $\mathcal{R}_0$. Two finite computations, $\mathcal{C}$ and $\mathcal{C}'$, are called *equivalent* if and only if they have the same answer substitution and the same mode of termination.

2.3.4 An Example for an FCP Program

An illustrative example of a simple FCP program represents the program *Quicksort*. In order to demonstrate how this program works, two example computations, a *succeeding* and a *suspending* computation, are presented below.

Both example computations apply the same process selection rule. Process selection is performed from left to right, according to the given ordering of processes within the resolvent. The most recently created processes are scheduled first, i.e. newly created processes are always added at the left-hand side of the resolvent. In order to identify the reducing program clause, a corresponding clause label is added to the transition rule type, e.g. $S_i \vdash_{REDUCE, <Clause\ Label>} S_j$.

Program Quicksort

% quicksort($InputList, OutputList$)

% From the integer list $InputList$ the
% corresponding sorted list $OutputList$ is computed.

$A0:$ $quicksort([X|Xs], Sorted) \leftarrow$
  $Xs \neq [\,] \,|$
  $partition(X, Xs?, S, L),$
  $quicksort(S?, S1),$
  $quicksort(L?, L1),$
  $append(S1?, [X|L1?], Sorted).$

$A1:$ $quicksort([X], [X]).$

$A2:$ $quicksort([\,], [\,]).$

$B0:$ $partition(A, [X|Xs], Smaller, [X|Larger]) \leftarrow$
  $A < X \mid partition(A, Xs?, Smaller, Larger).$

$B1:$ $partition(A, [X|Xs], [X|Smaller], Larger) \leftarrow$
  $A \geq X \mid partition(A, Xs?, Smaller, Larger).$

$B2:$ $partition(A, [\,], [\,], [\,]).$

$C0:$ $append([X|Xs], Ys, [X|Zs]) \leftarrow$
  $append(Xs?, Ys, Zs).$

$C1:$ $append([\,], Ys, Ys).$

Example Computation 1

States	Transitions

S_0 $\quad < quicksort([3, 1, 2], Sorted) \; ; \; \theta_0 = \emptyset > \; \vdash_{REDUCE, A0}$

S_1 $\quad < partition(3, [1, 2], S, L),$
$\qquad quicksort(S?, S1),$
$\qquad quicksort(L?, L1),$
$\qquad append(S1?, [3|L1?], Sorted) \; ; \; \theta_1 = \theta_0 > \; \vdash_{REDUCE, B1}$

S_2 $\quad < partition(3, [2], S', L),$
$\qquad quicksort([1|S'?], S1),$
$\qquad quicksort(L?, L1),$
$\qquad append(S1?, [3|L1?], Sorted) \; ;$
$\qquad \theta_2 = \theta_1 \circ \{S \to [1|S']\} > \; \vdash_{REDUCE, B1}$

S_3 $\quad < partition(3, [\,], S'', L),$
$\qquad quicksort([1, 2|S''?], S1),$
$\qquad quicksort(L?, L1),$
$\qquad append(S1?, [3|L1?], Sorted) \; ;$
$\qquad \theta_3 = \theta_2 \circ \{S' \to [2|S'']\} > \; \vdash_{REDUCE, B2}$

S_4 $\quad < quicksort([1, 2], S1),$
$\qquad quicksort([\,], L1),$
$\qquad append(S1?, [3|L1?], Sorted) \; ;$
$\qquad \theta_4 = \theta_3 \circ \{S'' \to [\,], L \to [\,]\} > \; \vdash_{REDUCE, A0}$

S_5 $\quad < partition(1, [2], S''', L'),$
$\qquad quicksort(S'''?, S1'),$
$\qquad quicksort(L'?, L1'),$
$\qquad append(S1'?, [1|L1'?], S1),$
$\qquad quicksort([\,], L1),$
$\qquad append(S1?, [3|L1?], Sorted) \; ; \; \theta_5 = \theta_4 > \; \vdash_{REDUCE, B0}$

States	*Transitions*

S_6 $< partition(1, [\,], S''', L''),$
     $quicksort(S'''?, S1'),$
     $quicksort([2|L''?], L1'),$
     $append(S1'?, [1|L1'?], S1),$
     $quicksort([\,], L1),$
     $append(S1?, [3|L1?], Sorted)\ ;$
     $\theta_6 = \theta_5 \circ \{L' \to [2|L'']\} > \ \vdash_{REDUCE, B2}$

S_7 $< quicksort([\,], S1'),$
     $quicksort([2], L1'),$
     $append(S1'?, [1|L1'?], S1),$
     $quicksort([\,], L1),$
     $append(S1?, [3|L1?], Sorted)\ ;$
     $\theta_7 = \theta_6 \circ \{S''' \to [\,], L'' \to [\,]\} > \ \vdash_{REDUCE, A2}$

S_8 $< quicksort([2], L1'),$
     $append([\,], [1|L1'?], S1),$
     $quicksort([\,], L1),$
     $append(S1?, [3|L1?], Sorted)\ ;$
     $\theta_8 = \theta_7 \circ \{S1' \to [\,]\} > \ \vdash_{REDUCE, A1}$

S_9 $< append([\,], [1, 2], , S1),$
     $quicksort([\,], L1),$
     $append(S1?, [3|L1?], Sorted)\ ;$
     $\theta_9 = \theta_8 \circ \{L1' \to [2]\} > \ \vdash_{REDUCE, A0}$

S_{10} $< quicksort([\,], L1),$
     $append([1, 2], [3|L1?], Sorted)\ ;$
     $\theta_{10} = \theta_9 \circ \{S1 \to [1, 2]\} > \ \vdash_{REDUCE, A0}$

S_{11} $< append([1, 2], [3], Sorted)\ ;$
     $\theta_{11} = \theta_{10} \circ \{L1 \to [\,]\} > \ \vdash_{REDUCE, C0}$

States	*Transitions*

S_{12} $\quad < append([2], [3], Sorted')$;

$\qquad \theta_{12} = \theta_{11} \circ \{Sorted \rightarrow [1|Sorted']\} > \vdash_{REDUCE,C0}$

S_{13} $\quad < append([\,], [3], Sorted'')$;

$\qquad \theta_{13} = \theta_{12} \circ \{Sorted' \rightarrow [2|Sorted'']\} > \vdash_{REDUCE,C1}$

S_{14} $\quad < true$; $\theta_{14} = \theta_{13} \circ \{Sorted'' \rightarrow [3]\} > \vdash_{SUCCEED}$

S_{15} $\quad < succeed$; $\theta_{15} = \theta_{14} >$

The observable behaviour of *Example Computation 1* corresponds to the
triple $< quicksort([3, 1, 2], Sorted)$; $\{Sorted \rightarrow [1, 2, 3]\}$; $succeed >$,
where the list $[1, 2, 3]$ represents the resulting answer substitution for the
input variable *Sorted*.

Example Computation 2

States	*Transitions*

S_0 $\quad < quicksort([1, 2|InputList], Sorted)$;

$\qquad \theta_0 = \emptyset > \vdash_{REDUCE,A0}$

S_1 $\quad < partition(1, [2|InputList?], S, L)$,

$\qquad quicksort(S?, S1)$,

$\qquad quicksort(L?, L1)$,

$\qquad append(S1?, [1|L1?], Sorted)$;

$\qquad \theta_1 = \theta_0 > \vdash_{REDUCE,B0}$

States	*Transitions*
S_2	$< partition(1, InputList?, S, L'),$
	$quicksort(S?, S1),$
	$quicksort([2|L'?], L1),$
	$append(S1?, [1|L1?], Sorted)\ ;$
	$\theta_2 = \theta_1 \circ \{L \rightarrow [2|L']\} > \ \vdash_{REDUCE, A0}$
S_3	$< partition(2, L'?, S', L''),$
	$quicksort(S'?, S1'),$
	$quicksort(L''?, L1'),$
	$append(S1'?, [2|L1'?], L1),$
	$partition(1, InputList?, S, L'),$
	$quicksort(S?, S1),$
	$append(S1?, [1|L1?], Sorted)\ ;$
	$\theta_3 = \theta_2 > \ \vdash_{SUSPEND}$
S_4	$< suspend\ ;\ \theta_4 = \theta_3 >$

The observable behaviour of *Example Computation 2* corresponds to the triple $< quicksort([1, 2|InputList], Sorted)\ ;\ \emptyset\ ;\ suspend >$. The computation suspends because of the fact that the input list is represented by a nonground term containing the variable *InputList* as an undefined tail list.

Chapter 3

Design of an Abstract FCP Machine

3.1 The Process Reduction Mechanism

The central operation in the execution of an FCP program is the process reduction.
Under the objective to increase efficiency when processing program code, main efforts
therefore concentrate on methods for improving the process reduction mechanism.

3.1.1 Complexity Issues

Efficiency, with respect to program execution time, is principally a matter of *clause
selection* and *clause evaluation* techniques. In general, the reduction of a process
necessitates to inspect several clauses of the related program procedure in order to find
an applicable clause enabling the reduction of that process. Each of these reduction
attempts may require to compute a number of resulting variable bindings. Whenever
an attempt to reduce a process by means of a selected clause fails, the obtained
variable bindings have to be *undone* as such an attempt must not leave any trace.
This kind of trial and error behaviour of the clause selection procedure is denoted as
shallow backtracking.

When a process reduction is delayed depending on the instantiation of one or more
read-only variables, clause selection may become even more costly. A process which
is scheduled several times, but cannot be reduced because of uninstantiated read-only
variables, requires to evaluate the whole program procedure more than once.

By application of an appropriate process suspension mechanism, the overall costs
for clause selection can be reduced significantly. A process thereby becomes suspended as soon as it tries to access a read-only variable. The same processes thereafter cannot be scheduled again as long as the read-only variables it is suspended on
remain uninstantiated. This way, the suspension mechanism avoids useless reduction

attempts and the overhead caused by *busy waiting* behaviour resulting otherwise. A concrete realization of the process suspension mechanism in an extended version for the distributed implementation is presented in some detail in Chapter 4.

Clause Selection. The execution of a program procedure $C^{p/k}$ consisting of the clauses $\{C_1^{p/k}, C_2^{p/k}, ..., C_r^{p/k}\}$ in an attempt to reduce a process $p(A_1, A_2, ..., A_k)$ typically divides into a number of single clause try operations; each of which corresponds to an attempt to compute the function $mgu_?(\ p(A_1, A_2, ..., A_k),\ C_i^{p/k})$ for some $i \in \{1, ..., r\}$ (cf. Section 2.3). Thereby, an individual clause try operation further separates into two basic suboperations being concerned with *head unification* and *guard evaluation*, respectively.

The abstract computational model of FCP assumes a nondeterministic clause selection mechanism, which would also allow to try all clauses belonging to the same program procedure in parallel. Any concrete realization of this model, at least when running on a uniprocessor machine, of course needs to specify some deterministic clause selection policy. A straightforward policy would be to select the clauses from the program procedure $C^{p/k}$ following the textual order they appear in the program. Somewhat more sophisticated approaches take into account the relative frequency a certain clause type contributes to a reduction. For instance, an iterative process may be reduced arbitrary often using iterative clauses; by means of a unit clause, however, it can be reduced only once. In any case, the implemented clause selection policy should appear transparent for the programmer.

Head Unification and Guard Evaluation. A considerable amount of time devoted to clause selection and clause evaluation is spent on unification. In an attempt to reduce a process $p(A_1, A_2, ..., A_k)$, the complexity of unification operations being performed in order to find a reducing clause among the clauses of $C^{p/k}$ basically depends on three facts:

     o the number of clauses r forming the program procedure $C^{p/k}$

     o the arity k and argument structures of each clause head

     o the structure and type of the process arguments $A_1, A_2, ..., A_k$

As we have already seen, FCP restricts the usage of guard predicates to a predefined set of primitive test operations. These test operations like arithmetic comparison, type checking, etc., do not require any complex subcomputations but can be computed immediately. For that reason, the evaluation of guard test predicates might also be considered as some kind of extended head unification. In principle, these operations could be embodied in a general unification algorithm as well.

Following this approach, an appropriate interleaving of head unification and guard evaluation would result in a simple but effective optimization. Due to the simplicity of guard testing, the compatibility of the process data state with a clause's guard should be checked concurrently to head unification. In particular, it is often possible to extract information relevant to guard testing prior to thorough unification of complex argument structures. This technique eliminates a great deal of superfluous unification overhead.

3.1.2 The Process Reduction-Cycle

Based on the regular structure of FCP program clauses and the typical data manipulation operations in processing them, each process reduction-cycle logically divides into two subsequent phases:

$$\underbrace{H \leftarrow G_1, G_2, ..., G_m}_{Phase1:\ Test\ Unification} \quad | \quad \underbrace{B_1, B_2, ..., B_n}_{Phase\ 2:\ Process\ Creation} \qquad (m, n \geq 0).$$

The first phase of a process reduction-cycle handles the unification of the process structure with the clause head as well as the evaluation of guard test predicates. Unification which is performed prior to processing the commit-operator serves as a means for controlling clause selection and process synchronization. At the same time, its purpose is to generate the input values from which the reduction operation computes the resulting output values.

The second phase spawns a set of parallel subprocesses replacing the process currently being reduced. Their argument structures are obtained as the result of unifying newly created variables with corresponding input values delivered by the preceding phase. In contrast to the preceding phase, any operation performed during Phase 2 is permanent.

3.2   The Abstract Machine Model

An efficient execution model for processing sequential Prolog code was proposed by Warren [Warren83] and is well-known as *Warren Abstract Machine (WAM)*. Meanwhile, this approach has established as a de facto standard for Prolog implementations. The basic idea is to define a mapping from Prolog clauses into a set of high-level instructions for a (virtual) sequential machine. This way, complex reduction operations are carried out by performing a sequence of specialized instructions.

As the granularity of operations decreases, they can be implemented much more efficiently on conventional computer architectures primarily being designed to execute imperative programming languages.

Much of the run-time overhead caused by application of universal unification algorithms can be eliminated through the above technique, which is denoted as *compilation of unification*. Utilizing the knowledge about the expected process argument structures, as it is encoded in the program clauses, general unification can often be replaced by a number of elementary but specialized test operations. The execution of a universal unification algorithm then is required only when checking data structures below top level argument structures already known at compile-time.

The original execution model of the WAM has been extended and improved in several ways. Especially, a further decrease of the granularity of the applied machine instructions as well as certain global optimizations have contributed to the efficiency of today's leading Prolog implementations. In fact, there is some evidence encouraged by recent results [Van Roy92] that future implementations of logic programming languages may achieve about the same efficiency as imperative language implementations. In combination with special hardware support, e.g. as provided by the KCM [Benker89], they may even overcome the imperative approach.

3.2.1   A Sequential FCP Machine

Most of the general optimization techniques embodied in the execution model of the WAM apply to concurrent logic programming languages as well. Compared to sequential Prolog, an abstract machine for a concurrent logic programming language like FCP is less complex, since it does not need the rather complex stack management facilities in order to handle backtracking, efficiently. On the other hand, a realization of dynamic data-flow synchronization without using busy waiting requires an additional process suspension mechanism. Finally, the concept of compiling unification has to be generalized and extended to read-only unification primitives.

An instruction set for a sequential FCP machine was first proposed by Houri and Shapiro in [Houri87]. In the design and implementation of our machine we use a different instruction set as well as different representations for FCP data structures. Nevertheless, there are certain similarities with respect to basic design concepts.

An overview on the architecture of our sequential FCP machine is contained in [Glaesser90c]; more detailed descriptions including the applied instruction set for the original and an improved version of this machine are presented in [Lehrenfeld90] and [Hannesen91], respectively. With the objective to obtain a performance estimation for a hardware realization, a further improved machine architecture was investigated in [Grunzig92].

3.2.2 Functional Machine Architecture

Without going far into the details, some basic issues concerning the functional architecture of the sequential FCP machine, especially the representation of dynamic data structures, are briefly discussed.

Representation of FCP Data. Within the sequential FCP machine, there are three storage areas for representing dynamically created data objects, as illustrated in Figure 3.1. Beside the main storage area, which is called the machine's *Heap*, there is a *Trail-Stack*, and a set of registers $X_1, ..., X_n$. The overall organization of the machine is that of a *tagged architecture*. An individual machine word consists of a *tag field* and a *data field:* $< tag >< data >$. The tag field identifies the type of the stored data item, whereas the data field contains the corresponding data value. Similar to Prolog data structures, the basic FCP data type is the *logical term*. Indeed, the data structures of Prolog and FCP are identical, except for the fact that Prolog does not have read-only variables.

According to the general distinction between *atomic* terms (e.g. integer, character, or variables) and *compound* terms (e.g. lists or tuples), an FCP data object may either be atomic or it may be *compound*. Atomic data objects are always represented by a single machine word. In order to represent a compound data object consisting of n atomic data objects, at least $n + 1$ machine words are required.

In addition to ordinary FCP data objects, a number of special *pointer objects* is used for the machine layer representation of data. Examples of pointer objects are list pointer *(List)*, tuple pointer *(Tuple)*, and pointer to variables *(Ref)*. Each of these pointer objects is represented by a single machine word and has the general form $< tag >< pointer >$, where *tag* identifies the pointer type and *pointer* refers to a machine address.

Pointer objects are used in order to represent compound data objects. Lists and tuples are represented in the following form $< Header >< Item\ 1 > ... < Item\ n >$. The *Header* of a list or tuple is a corresponding pointer object referring to the first address of a list of consecutive machine words which contain the actual tuple or list items. Each item *Item i* $(1 \leq i \leq n)$ again may represent a compound or an atomic data object. The internal representation of a tuple $f(T_1, ..., T_k)$ has the form $(f, T_1, ..., T_k)$, where the arity of the tuple is encoded in the tuple pointer. In order to recognize the end of a list, a special *EOL* marker is encoded in the tag field of the last list element. (Figure 3.1 gives an example.)

In addition to the usage of pointer objects as described above, pointer objects of the type *Ref* serve as a mean to describe variable bindings. For example, if the variable X becomes bound to another variable Y, X is replaced by a pointer object of the form $< Ref >< address(Y) >$.

Beside the dynamic data structures mentioned so far, the *Heap* also contains the FCP process structures as well as the data structures required to handle process suspensions. In fact, all these data structures are represented as tuples, which are dynamically *created* and *deleted*. The machine representation of a process structure consisting of k arguments corresponds to a tuple of arity $k + 2$. The additional entry thereby identifies the program procedure belonging to the process.

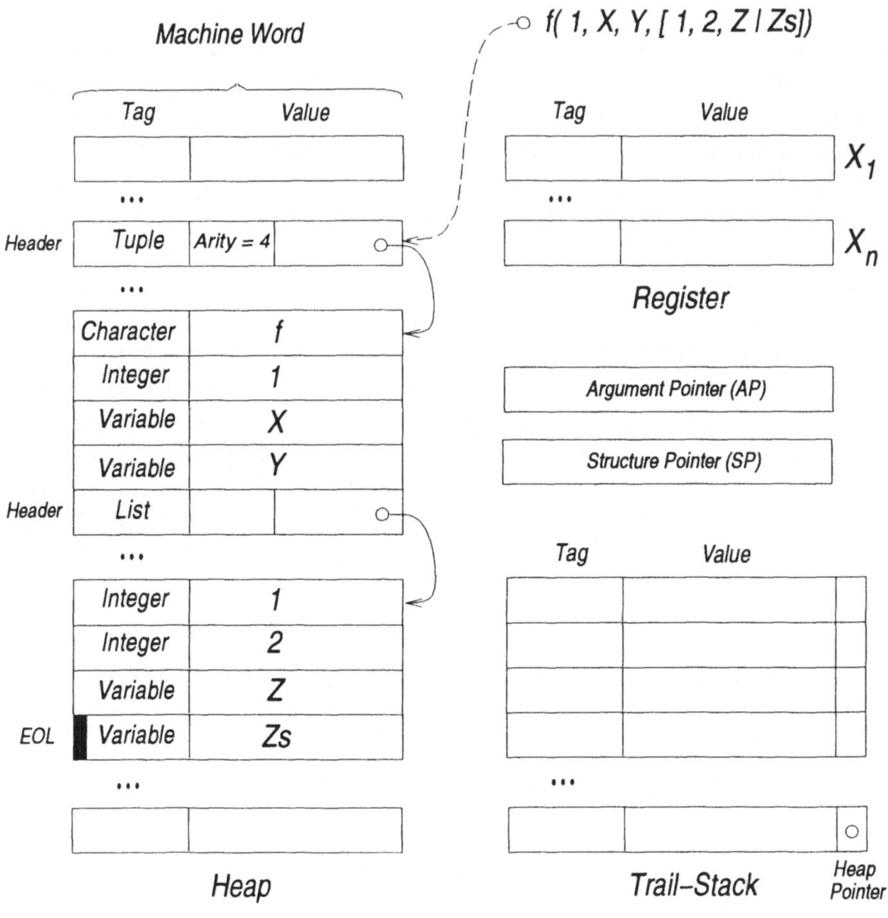

Figure 3.1: Organization of the Sequential FCP Machine

Operational Behaviour of the Sequential Machine. Within each reduction cycle, the machine attempts to reduce an FCP process by means of a selected clause from the related program procedure. One by one, the process arguments are unified with the arguments of the clause.

While the argument structures of the clause are encoded in the machine instructions of the program procedure, the heap addresses corresponding to the process arguments are referred by means of the argument pointer AP. The individual substructures of a complex argument structure are referred by the structure pointer SP.

Intermediate unification results corresponding to bindings of clause variables are stored in the registers $X_1, ..., X_n$. Intermediate unification results corresponding to bindings of process variables, however, are stored in the *Heap*. This means that process variables are overwritten. In order to be able to restore the old state of the *Heap*, in case that a reduction attempt fails, the *Trail-Stack* is used. Before a variable is modified, the variable as well as the address of this variable are saved on the *Trail-Stack*. Depending on the outcome of a reduction attempt, the contents of the *Trail-Stack* is either ignored or copied back to the *Heap*.

3.2.3   Process Scheduling

A central operation in the execution-cycle of the sequential FCP machine is process scheduling (see Figure 3.2). Although a process suspension mechanism may already eliminate much of the overhead caused by useless reduction attempts, the choice of the process scheduling policy has a strong impact on the overall machine performance. In order to characterize this relationship, it is helpful to identify various kinds of *process reduction costs*. With respect to a particular application program $\mathcal{P}$ and a particular realization of a sequential machine, these costs are:

$c_{Schedule}$: *Mean costs for scheduling an individual process*

c_{Commit}: *Mean costs for determining an applicable clause in a successful reduction attempt*

$c_{Suspend}$: *Mean costs for clause evaluation in a suspending reduction attempt*

c_{Extra}: *Mean costs caused by extra handling required for the process suspension mechanism*

The scheduling costs $c_{Schedule}$ especially depend on the complexity of applied scheduling policy. The expression c_{Extra}, in particular, includes the costs for suspending a process on one or more variables and for waking up the process again when the variables become instantiated.

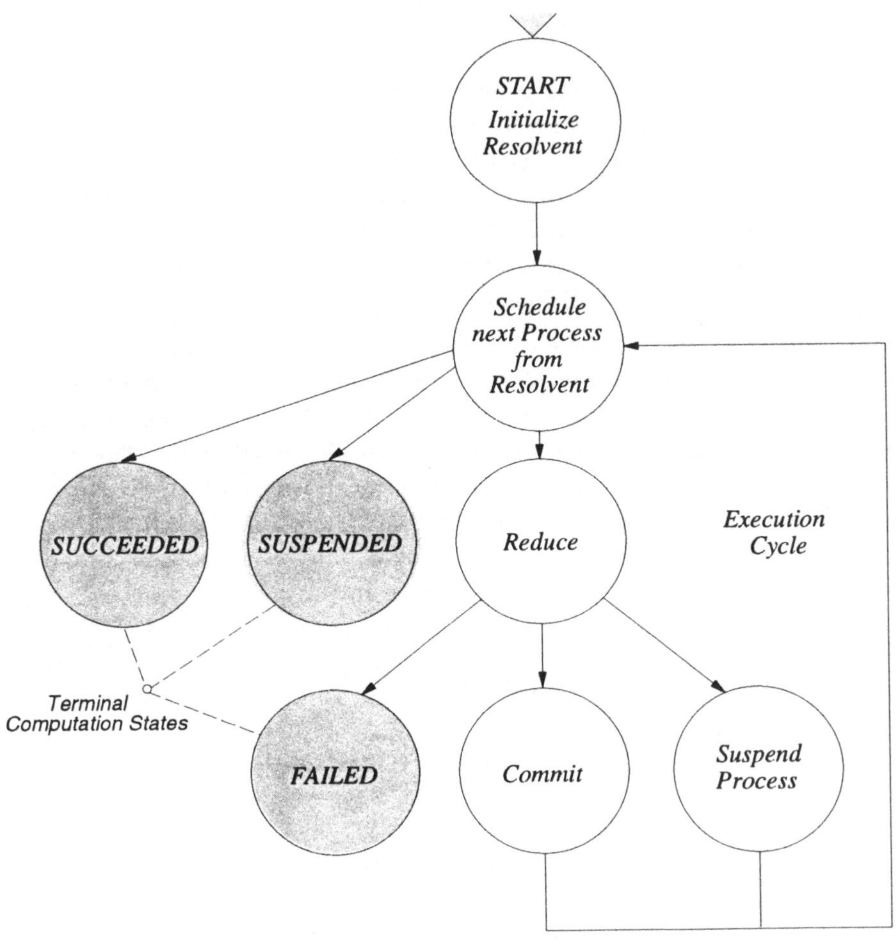

Figure 3.2: State Diagram of the Sequential FCP Machine

In a finite computation of the program $\mathcal{P}$ on some given input data set $\mathcal{I}$, an estimation of the total reduction costs $c_{Reductions}$ depending on the implemented deterministic scheduling S are given by:

$$
\begin{aligned}
c_{Reductions}(S) = \; & n(c_{Schedule} + c_{Commit}) + \\
& k(c_{Schedule} + c_{Suspend} + c_{Extra}), \quad k \geq 0, \; n > 0,
\end{aligned}
$$

where n denotes the number of *successful* reduction operations and k denotes the number of reduction attempts resulting in a process *suspension*.

A main issue, when comparing different deterministic scheduling policies, is the *locality* of computation with regard to the traversal of the computation tree. At the same time, another dimension of scheduling, which is *fairness*, has also to be taken into consideration. The restrictions implied by fairness requirements, of course, depend on the underlying definition of fairness and may be rather different for different applications. However, there usually arises a conflict between the goals of ensuring maximum fairness and achieving maximum performance.

A general comparison of two alternative scheduling policies, *Bounded Depth First Search (BDFS)* [Shapiro86] and *First In Last Out (FILO)*, is presented in [Grunzig92]. Using several standard bench mark programs, the obtained experimental results demonstrate that *FILO* is superior to *BDFS* in almost all cases. This is due to the locality of computation, which is preserved much better when using *FILO*. On the other hand, a naive *FILO* policy does not comply with fairness rules. Hence, an integrated solution combining *FILO* with *BDFS* might be a successful approach. Unfortunately, this has not yet been investigated further.

Chapter 4

Concepts for a Distributed Implementation

4.1   Abstract System Architecture

In order to point out substantial design issues for the parallel FCP machine, this section introduces the underlying abstract machine architecture in terms of a generic model. The resulting *abstract machine specification* provides a general framework within which the various techniques used for exploiting and controlling parallelism are discussed in the sections 4.2 to 4.9. In fact, this specification already reflects the basic features of the parallel machine, as far as they concern the *distributed reduction algorithm* and *dynamic work load balancing*. Important design issues refer to the applied parallelization concept, the resulting scalability features, communication and synchronization, and distributed data representation. How to implement the abstract architecture on a real target architecture, with respect to an implementation on Transputer networks, will be considered in Chapter 5.

The whole parallel machine is organized as a network consisting of $n+1$ asynchronously operating processing elements. According to the function they realize, the processing elements separate into n so-called *reduction unit* $RU_0, ..., RU_{n-1}$ ($n \geq 1$), and a special *host unit*. While the uniformly constructed reduction units represent the machine's basic building blocks, the additional host unit takes the role of a central supervisor. Via a *communication network (CN)*, built up from bidirectional point-to-point links, the reduction units as well as the host unit are clustered as shown in Figure 4.1.

An individual reduction unit essentially provides the functionality of a sequentially operating FCP machine, but has some extra capabilities to cooperate with other reduction units. Its core component executes compiled FCP code on a private local memory. Based on the WAM execution model, this core component is realized almost the same way as described in Chapter 3 for a purely sequential FCP machine.

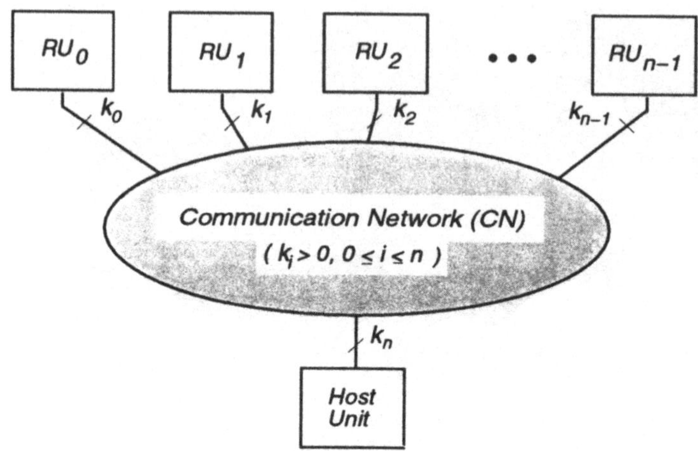

Figure 4.1: Abstract Architecture of the Parallel FCP Machine

However, there are two essential differences between a sequential and a parallel version of an abstract FCP machine. Beside a slightly extended unification algorithm the parallel version, of course, requires a communication interface. The modified unification algorithm enables the machine to handle *remote data objects* in addition to local ones. For that purpose a new pointer type, the *cross environment reference (XER)*, has to be introduced. The communication interface allows to perform interactions with remote units as required when referring to remote data objects. It is implemented using *message-passing* communication and synchronization primitives.

4.1.1 Parallelization

The parallel FCP machine offers parallelization at the process reduction layer, i.e. the basic *unit of parallelism* is the FCP process. A given number of reduction units $RU_0, ..., RU_{n-1}$ cooperatively perform a distributed computation by interleaving concurrent reduction operations. Each reduction unit RU_i $(0 \leq i \leq n-1)$ therefore runs an identical copy $\mathcal{P}_i$ of the same FCP program $\mathcal{P}$.

Parallelization is achieved by partitioning the *global* resolvent $\mathcal{R}$ into n subsets representing the *local* subresolvents $\mathcal{R}_0, ..., \mathcal{R}_{n-1}$; each of which is assigned to one of the reduction units $RU_0, ..., RU_{n-1}$. This way, the global computation becomes broken up into a corresponding number of local subcomputations. The resulting mapping of the global process network onto the network of reduction units executing it reflects the distribution of work load over the parallel machine.

Dynamic Load Balancing. While performing a distributed computation, the size of the local subresolvents and thereby also the balancing of work load dynamically changes. By application of a dynamic load balancing algorithm, the work load distribution automatically adapts to continuously changing load situations. This corresponds to a dynamic partitioning of the global process network, which is achieved by reorganizing the local subresolvents. Processes are therefore enabled to migrate between reduction units.

Depending on frequently computed local load indices, the distributed load balancing algorithm initiates and controls process migration according to the implemented dynamic load balancing policy. All activities related to dynamic load balancing are totally distributed to the reduction units. Furthermore, theses activities are carried out in such a way that load balancing is transparent at the application language layer.

In order to initiate a distributed computation, the host unit sends the input processes which represent the initial process network to RU_0. Starting from this initial configuration RU_0 becomes busy while all the other reduction units $\mathcal{R}_1, ..., \mathcal{R}_{n-1}$ are still idle. Now, migration of processes effects the work load to be incrementally distributed over the network.

Which processes migrate to which reduction units and under what conditions they do this, merely depends on the particular load balancing policy being implemented. Though, there is a general restriction. Independently of the load balancing policy, only active processes may migrate between reduction units.

The Parallelization Model. The described computation scheme corresponds to an *AND-parallel* execution model ([Conery87],[Kurfess91]), i.e. the concurrently executed subcomputations operate on a common set of global variables. As a consequence, there usually result interdependencies between individual subcomputations that may reduce the effective degree of utilizable parallelism. However, this seems to be a reasonable approach because the number of processes in real applications is significantly higher than the number of reduction units. The resulting parallelization behaviour therefore is *coarse grain* processing rather than *fine grain* processing [May90].

With respect to general design concepts of distributed systems and parallel machine architectures, the way a distributed computation of the parallel FCP machine is organized reflects a common approach. Though all units run the same program, each unit effectively executes a different stream of instructions on a different stream of data. In other words, the overall machine design follows the *multiple instruction stream-multiple data stream (MIMD)* scheme according to Flynn's taxonomy in [Flynn66].

4.1.2 Scalability

A main objective in the design of the parallel machine is to achieve maximum scalability. Different application programs may show rather different parallelization behaviour. Hence, it is desirable to have a parallel machine which easily scales with respect to the extent of parallelization it provides. The proposed parallel machine architecture offers maximum scalability by varying the size of the network within a wide range. In fact, the total number n of reduction units that build up the machine network remains an almost free parameter.

A minimal configuration is given by a network consisting of a single reduction unit in combination with the host unit ($n = 1$). Such a configuration actually realizes a sequential FCP machine. As there is just a single operating reduction unit keeping all data local, no additional overhead for communication or synchronization occurs.

The maximum value for n, however, represents a limitation placed by the underlying hardware system resources. It merely depends on the number of processors that are available. Similarly, the maximum number of links interconnecting an individual reduction unit or the host unit with the communication network is determined by the hardware system, too.

Arbitrary Topologies.   The parallel machine does not require any particular network topology, such as a hypercube, a torus, etc., but it dynamically adapts to arbitrary network configurations. When being started on some initially unknown network, the network is explored under control of the host unit.

To each processor belonging to the network a reduction unit together with a unique *unit identifier* and a local *routing vector* is assigned.   The necessary topology information is extracted from the network using an *all-pair shortest-path* distributed algorithm [Ramme90]. This allows to generate a global routing matrix from which the local routing vectors then are derived.

Parallel Machine Organization.   The overall organization of the parallel FCP machine, as shown in Figure 4.2, is such that any effective parallelization is determined and carried out at run-time. In fact, this means that the programmer need not to know how many processors are involved in the execution of a program and how the global process network has to be mapped onto these processors. The same application program may run on a uniprocessor configuration as well as on arbitrary large networks with any nummer of processors. Consequently, the compiler needs also not to know whether the code it produces is to be executed on a sequential or parallel machine. On the other hand, it still remains possible to extract helpful information from a program at compile-time in order to support or improve parallelization whenever the program is executed by more than one processor.

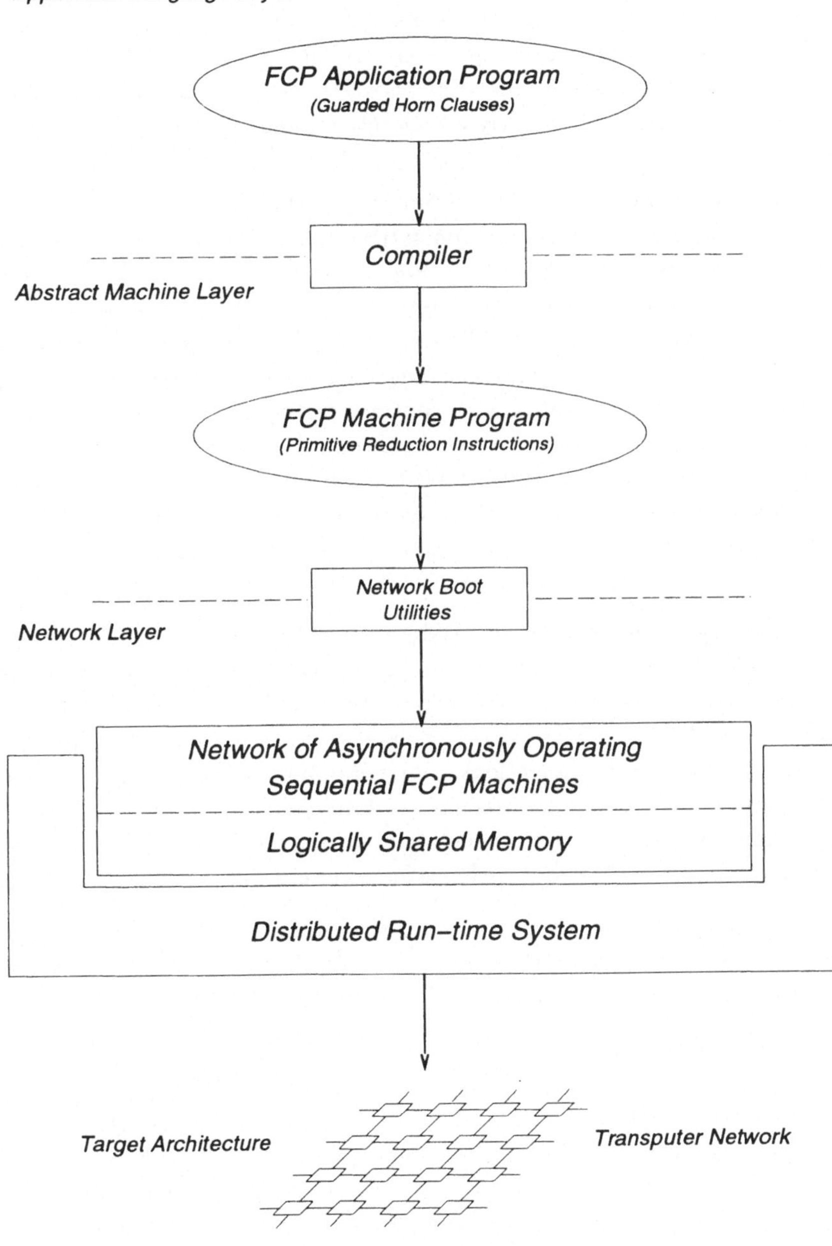

Figure 4.2: Organization of the Parallel FCP Machine

4.1.3   Communication and Synchronization

Attributes concerning *communication* and *synchronization* refer to most important issues in the design of distributed systems. Moreover, the basic questions of how to realize efficient communication between parallel units, especially those residing on different processors, and how to meet the necessary synchronization demands ensuring that these units cooperate properly are closely related.

Interprocess Communication.   A commonly used abstraction when addressing aspects of communication and synchronization in parallel or distributed systems is expressed by the term *interprocess communication (IPC)*. With the objective to implement a concurrent logic programming language on a message-passing multi-Transputer system, one is concerned with two basically different models of IPC.

The IPC model of the application language, which is based on communication through shared logical variables, is in contrast to the message-passing model of the Transputer hardware architecture. In addition, the shared variable model applies *asynchronous communication* while the IPC model of the Transputer, which is defined by CSP [Hoare78, May85], merely offers *synchronous communication* (cf. Figure 4.3). Basic IPC primitives of the application language, therefore, have to be efficiently mapped onto corresponding primitives for the Transputer.

Communication through shared variables not only is a very natural but also is the most frequently used method for IPC in various kinds of computing systems. With multiprocessor systems, however, this approach has usually been taken for tightly coupled systems [1], only, while loosely coupled systems more often apply some form of message-passing. Both realizations, in general, have their own advantages as well as disadvantages concerning addressing, flexibility, and synchronization features. At the same time, shared variables on one hand and message-passing on the other hand represent two opposite extremes within a spectrum of different solutions to IPC.

Recently, several alternative approaches that might be settled in between shared variables and message-passing have been introduced. An overview and a comparison thereof is presented in [Bal88, Bal89]. Though the related systems differ widely with regard to their semantics and the applied mechanisms for addressing and synchronization, an important common aspect is *replication of data.*

The objective in having multiple copies of the same data object residing on different processors, actually, is to decrease *access time* rather than to increase *availability.* Unfortunately, replication of data within a distributed system introduces a new problem. Almost the same problem, for example, also occurs with distributed data base

---

[1]The distinction between *tightly coupled* and *loosely coupled* multiprocessor systems, traditionally, corresponds to a classification into shared-memory and nonshared-memory architectures. The later form sometimes is also referred to as distributed systems.

systems or even with classical tightly coupled multiprocessor systems when using multiple caches. In the former system class it is known as *data consistency problem;* in the later one it is known as *data coherence problem.*

For both kinds of systems the problem has been studied, extensively, and a number of solutions already exist [Bernstein81, van de Goor89].

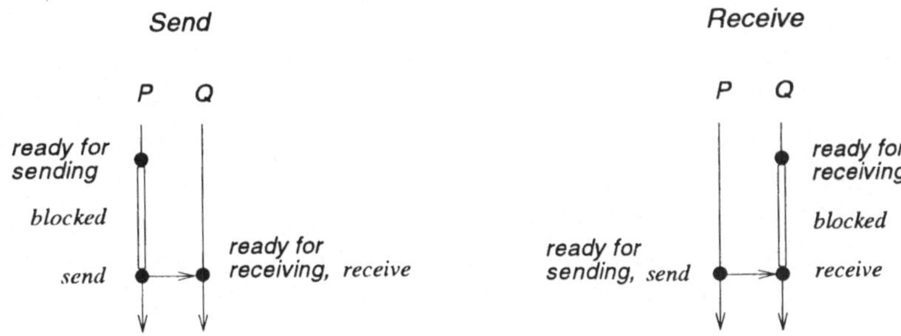

a) Synchronous Communication Behaviour

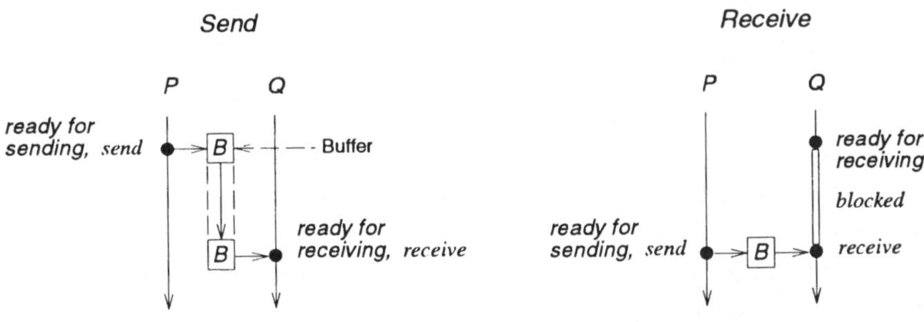

b) Asynchronous Communication Behaviour

Figure 4.3: Basic Models of Interprocess Communication Behaviour

An important observation is that these problems become relatively easy to handle as long as the replication of data restricts on *immutable* data. The complexity of update operations significantly reduces if data objects once they have been copied need not to be changed any more. In the particular case of a concurrent logic programming language which is to be implemented on a on a loosely coupled multiprocessor system, this concept naturally matches with the *single-assignment feature* of logic variables.

The Message-Passing System.   For the abstract architecture of the parallel FCP machine it is convenient to assume that any interaction between reduction units rests upon asynchronous message-passing communication. Moreover, the particular message-passing model used here, in contrast to other ones – e.g. as specified in [Bal89], applies *asynchronous point-to-point communication*. Using the unique unit identifiers in combination with local routing matrices, messages between units not being direct neighbours are always forwarded following a shortest path.

The message-passing mechanism as it is embedded in the machine's distributed run-time system guarantees to be both, *reliable* as well as *order-preserving*. To be reliable means that messages cannot get lost or miss their receiver as a result of transmission failures.

Because of the asynchronous communication behaviour, there may appear a number of pending messages between a sending unit RU_i and a receiving unit RU_j. That is, messages that have already been sent by RU_i but either have not yet been received or not yet been processed by RU_j. As the message-passing mechanism is order-preserving, it always guarantees that pending messages are not mixed up. At the receiver unit they are processed in exactly the same order as they have left the sender unit.

4.2   Distributed Data Representation

The parallel machine architecture might in principal be classified as a *physically distributed system* with *logically shared memory*. While the concept of a physically distributed system reflects the view at the parallel machine layer, the concept of a logically shared memory corresponds to the view at the application language layer. In the implementation of the parallel machine the logically shared memory therefore has to be simulated in such a way that any physical distribution of data appears transparent to the application programs.

4.2.1   Data Representation at the Application Layer

For a reduction unit's sequential machine component data objects always appear to be local, i.e. accessible through a reference into the local memory. In case that the physical representation of a requested data object does not reside on the local processor, evaluation of the local reference yields the value of a so-called *cross environment reference (XER)*. The XER then identifies the actual location of the data object on a remote processor either immediately or via a chain of related XERs. By the term *cross environment reference* a reference pointing from a heap location inside the local environment of some processor P_i to a related heap location inside the local environment of another processor P_j is denoted.

Whenever a data access operation encounters a XER, regardless whether this is an attempt to read or to write a data item, an explicit data request message must be generated in order to get the remote data object local. As far as the sequential machine component of the requesting reduction unit is concerned, access to remote data objects compared to local ones just cause an increase of the delay time.

In general, the exact duration of this delay cannot be determined a priori, as it depends on a number of variables; e.g. the communication load on the network and the response time of the target reduction unit. In order to avoid busy waiting, the process which refers to the XER is suspended on this XER until the requested data object becomes available. Thus, the resulting process behaviour when encountering a remote data object is exactly the same as the behaviour effected by read-only variables.

4.2.2 Data Representation at the Machine Layer

The machine layer representation of FCP data distinguishes between two basically different types of data objects; these are *local data objects* and *global data objects*. In contrast to local ones, global data objects represent globally shared data and need to be accessible by more than one reduction unit at a time. Whether a data object is to be regarded as local or global does not depend on the particular FCP data type but depends on certain run-time conditions.

Distributed representation of globally shared data is based on the concept of a *virtual global memory*. The underlying global address space results from combining all the local address spaces of the individual reduction units. A global address x then has the form $x = < i, j > (0 \leq i \leq n - 1)$, where i identifies the reduction unit and j a corresponding local heap address.

In principal, there is a unique global address space including both, global data as well as local data. Though, when considering the realization of the distributed memory, it appears much more like a two-level *memory hierarchy* than a homogeneous global memory. A number of local submemories are built upon a common global memory (Figure 4.4). Since local data objects do not require any special treatment, it is convenient to regard the common global memory as the component maintaining global data, while all local data is distributed over the local submemories.

The following three subsections concentrate on the realization of the virtual global memory. In order to get the right understanding of the described solution, it is important to realize how the *single assignment nature* of logical variables is preserved in the execution model of concurrent logic programming languages. As no backtracking is applied, any variable bindings that are computed during a reduction operation become permanent as soon as the commit-operator has been processed.

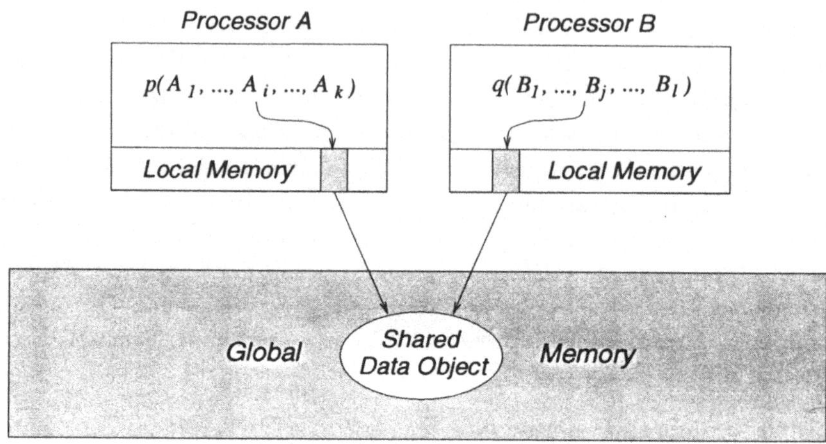

Figure 4.4: Representation of Globally Shared Data

In fact, this allows to consider non-variable terms as **immutable data**. For that reason, it is easily possible to replicate and distribute non-variable terms without riscing complex update operations. When a variable becomes instantiated with a non-variable term, the resulting data structure can be copied to any reduction unit which has been referring to the former variable. The same handling does even apply to non-ground compound terms, i.e. terms that consist of structures itself containing variables. In the copied structures all embedded variables are replaced by corresponding XERs.

4.2.3 Representation of Non-Variable Terms

Data redundance as obtained by replication of immutable data provides an appropriate means to reduce the access time to global data structures within a distributed environment. Nevertheless, when dealing with complex structures the resulting communication costs must also be taken into account. In addition to pure communication costs, this also includes costs for packing and unpacking structures. As far as the parallel FCP machine is concerned, these costs can be significantly reduced when using data distribution by means of *demand-driven structure copying* [Treleaven82].

Complex data structures are usually constructed from a number of recursively nested substructures. Instead of replicating and distributing the whole structures in one step, it is more efficient to perform remote copy operations, incrementally, on demand. For the unification algorithm to determine whether two deeply nested structures are unifiable or not, it is often sufficient to know the type of the top level argument structures. This aspect has also been investigated in [Taylor89].

Based on this insight, the communication costs caused by data distribution can be reduced when limiting the maximum number of levels k that are copied in one step. In case that the depth of a structure exceeds the value k, the remaining substructures are cut off and copied on explicit demand, only. Although, this technique provides a useful optimization, there still remains a problem. The optimal value k depends on the application.

Incremental structure copying, of course, requires to represent and handle *incomplete data structures*. For the described parallel machine architecture, this problem can easily be solved by means of XERs. Missing substructures within a partially copied data structure, i.e. parts that have been cut off, are represented by corresponding XERs. If necessary, evaluation of such XERs identifies the locations of the missing parts in order to perform further requests. Partial replacements within replicated data structures are always applicable for any complete substructure since a XER may refer to any kind of data object.

4.2.4 Representation of Logical Variables

Data replication as used for non-variable terms effectively realizes a *multiple-path multiple-data system (MPMD)* [van de Goor89]. The same global data item may be accessed through multiple pathes via different reduction units.

For the representation of globally shared variables, however, the usage of a MPMD scheme has a substantial drawback. The existence of multiple access pathes would considerably increase the control overhead to be paid in order to maintain data consistency and to avoid access conflicts. This drawback becomes eliminated when using a classical *single-path single-data (SPSD)* organization [van de Goor89], where each data item is under exclusive control of a single reduction unit [2].

Distributed variable representation as specified below, effectively realizes a SPSD solution with respect to data access control. At the same time, both concepts, the SPSD scheme as well as the MPMD scheme, can be efficiently combined into a common model. Because of the clear separation between immutable data on one hand and logical variables on the other hand, this seems to be a reasonable approach.

The Distributed Representation Scheme. The distributed representation scheme of (globally) shared logical variables has to be considered in combination with a set of rules controlling its dynamic evaluation. The representation scheme together with the evaluation rules must ensure a number of features which are important for the realization of the distributed reduction algorithm:

---

[2] The same control strategy has also been applied in [Taylor87]. However, there are fundamental differences between the underlying schemes of distributed variable representation. (A comparison is deferred to subsequent sections.)

o Whenever two or more reduction units attempt to modify the same variable, concurrently, **mutual exclusive write access** must be guaranteed.

o The operation of binding a write-enabled variable with a non-variable term or another variable is performed as an **atomic action.**

o When a write-enabled variable becomes instantiated, it must be possible to identify all related read-only variables.

To perform variable bindings as atomic action means, that any attempt to affect a variable during head unification or guard evaluation will have one of two definite results; the attempt either fails without leaving any trace or, in case it succeeds, any occurrence of this variable, independent of its location, becomes affected in the same way.

Formally, the distributed representation scheme of a logical variable 'X' corresponds to a *directed acyclic graph (DAG)* $G_X = (V, E, attr)$, which is extended by additional node attributes. The vertex set V together with the set of directed edges E define the distributed structure of X.

To each vertex $v \in V$ an attribute identifying the particular vertex type is assigned using the function $attr : V \longrightarrow \{local,\ remote,\ read\text{-}only\}$.

Definition. Let $G_X = (V, E, attr)$ specify the *distributed representation scheme* of a globally shared variable X. For some subset of vertices $\{v_1, v_2, ..., v_k\} \subseteq V$ let $v_1 \overset{x}{\leadsto} v_k$, $x \in \{remote, read\text{-}only\}$, denote a path $v_1 \rightarrow v_2 \rightarrow ... \rightarrow v_k$ in G_X, such that the following property holds:

$$(\forall i)\ 1 \leq i \leq k - 1 :\ (v_i, v_{i+1}) \in E \ and \ attr(v_i) = attr(v_k) = x.$$

A legal representation scheme must satisfy the three conditions defined below, where $\hat{u}$ denotes a particular vertex contained in V and $V' = V - \{\hat{u}\}$:

1. $attr(\hat{u}) = local \ \wedge \ \forall(v \in V') : attr(v) \neq local$

2. $\forall(u \in V') :$
   $attr(u) = remote \Rightarrow \exists(u' \in V) : u \overset{remote}{\leadsto} u' \rightarrow \hat{u} \ is \ a \ path \ in \ G_X.$

3. $\forall(v \in V') :$
   $attr(v) = read\text{-}only \Rightarrow \exists(v' \in V') : \hat{u} \rightarrow v' \overset{read\text{-}only}{\leadsto} v \ is \ a \ path$
   $in \ G_X.$

Within the distributed variable representation scheme G_X the unique *local* vertex $\hat{u}$ identifies the current location of the physical variable representation. Initially, this is the location where the variable is created. If there exist additional writable occurrences of X which are residing on remote reduction units, these are represented as XERs pointing to the physical variable location.

More precisely, a XER may either point to the physical variable location, immediately, or it may point to another XER also related to X. In the latter case, recursive evaluation of a chain of XERs eventually identifies the physical variable location. In the attributed graph G_X the XERs are reflected by *remote* vertices.

The third vertex type, the *read-only* vertices, correspond to read-only occurrences $X?$ of the variable X. Though, read-only occurrences are also represented by some kind of cross environment references, similarly to *remote* vertices, they require special handling. Together with the *remote* vertex $\hat{u}$, the *read-only* vertices form a directed subtree in G_X, where $\hat{u}$ represents the root of this subtree (Figure 4.5).

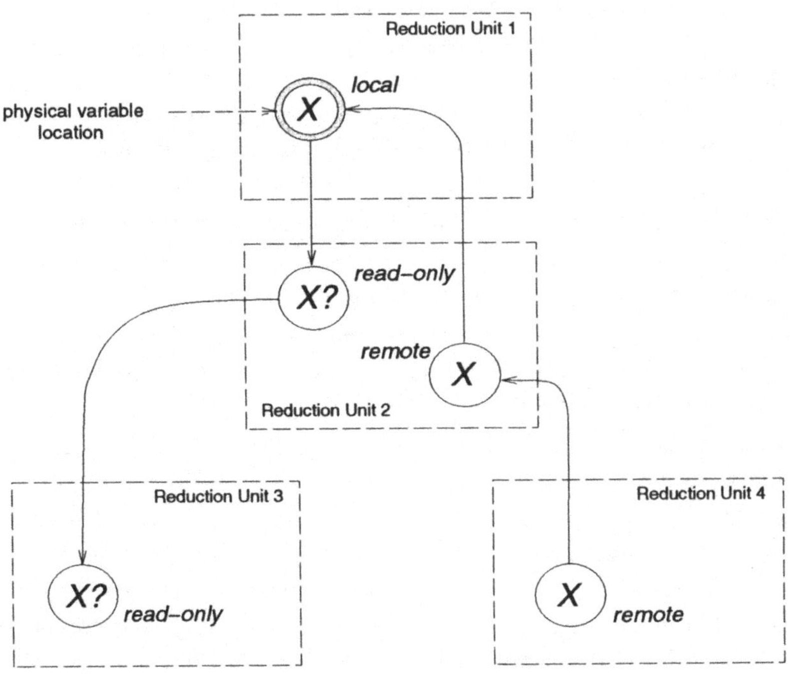

Figure 4.5: Distributed Variable Representation

With respect to the distributed representation of a globally shared variable X, as defined by G_X, the following notion applies to reduction units:

> **Definition.** The reduction unit keeping the writable variable occurrence corresponding to the *local* vertex $\hat{u}$ is called the *variable owner;* whereas reduction units that hold writable occurrences corresponding to *remote* vertices are denoted as *variable members*.

Using the owner/member relationship between reduction units sharing a common variable X, the variable owner and the variable members respectively operate with different access modes. Immediate access to the variable X in order to bind it with another variable or to instantiate it to a non-variable term is permitted to the variable owner, exclusively. For any reduction unit other than the variable owner an attempt to access X effects an immediate suspension of the corresponding process.

4.2.5 Representation of Process Structures

The dynamic load balancing mechanism requires that processes are able to migrate between reduction units. In order to realize such a dynamic re-allocation of processes to processors, there must be a suitable scheme for *distributed process representation*. When the location of a process $P = p(A_1, A_2, ..., A_k)$ $(k \geq 1)$ changes, as P migrates from the reduction unit RU_r where the process has been created to another reduction unit RU_s $(s \neq r)$, then the process arguments $A_1, A_2, ..., A_k$ have to be replicated accordingly. In general, it is not possible to delete the process arguments at their old locations at RU_r since there may exist other processes still referring to them.

Replication of process arguments structures basically obeys the same rules which also determine the distributed representation of individual variables and nonvariable terms. The operation is carried out straightforward whenever process arguments consist of ground terms, only. Beside ordinary ground terms this also includes terms containing $XERs$ but no variables.

However, in case of arguments that are either represented by variables or terms containing variables, the situation becomes somewhat more complicated. The way these variables are handled has a considerable impact on the global behaviour of the distributed reduction algorithm. At the same time, it also affects the communication and synchronization costs due to dynamic load balancing. Essentially, there are two alternative solutions for representing process argument variables; both of which have their own advantages as well as disadvantages. They are discussed below.

Alternative I. Figure 4.6 and Figure 4.7 illustrate the first approach. When process P migrates from RU_r to RU_s, the related *write-enabled* variables remain

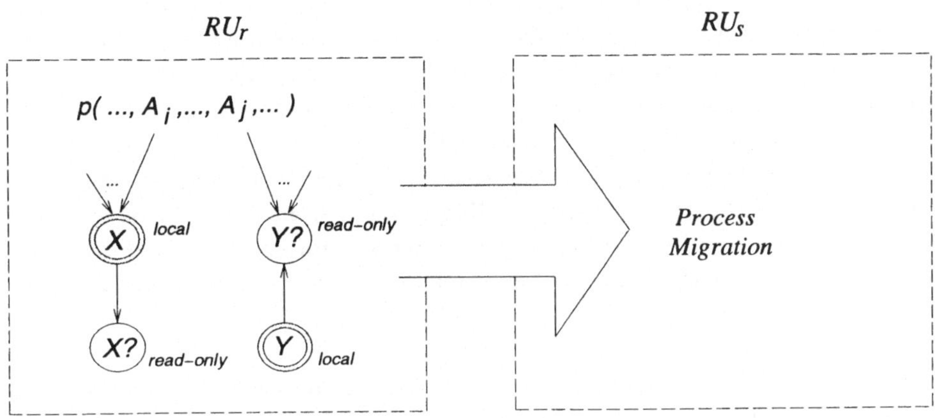

Figure 4.6: Initial Process Configuration

at their original locations at RU_r, while the corresponding variable occurrences at RU_s are represented by *XERs*. *Read-only* variables, in contrast to write-enabled variables, are simply replicated at RU_s but have to be attached to their write-enabled counterparts at RU_r (Figure 4.7).

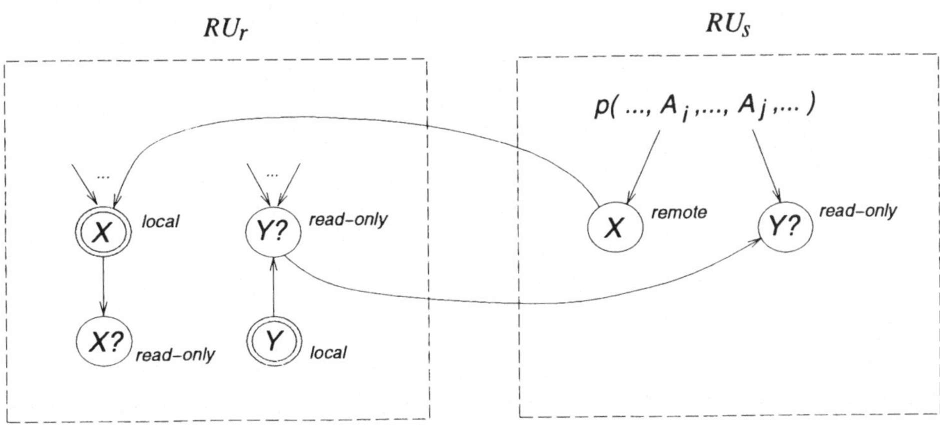

Figure 4.7: Distributed Representation of Processes (Alternative I)

Though such a representation provides an appropriate solution with respect to elementary synchronization demands, it yet has a decisive drawback. Write-enabled variables usually represent output channels of a process. Therefore, there is some evidents that process P which is going to migrate from RU_r to RU_s will attempt to bind these variables in the environment of RU_s.

For most application programs it is even typical that there is just a single process attempting to bind a certain variable. Hence, a representation scheme as shown in Figure 4.7 would result in a dramatical increase of communication and synchronization costs because of the high number of remote access operations that will become necessary.

Alternative II. Figure 4.8 illustrates the second approach of how to realize process migration. It shows the resulting configuration which is obtained from the same initial configuration that has been assumed in Figure 4.6. Instead of replacing the write-enabled variables within the replicated process arguments at RU_s, the original variable locations at RU_r are converted into $XERs$. As a consequence, the variables remain local to the process which is most likely to bind them. Read-only variables are treated as before.

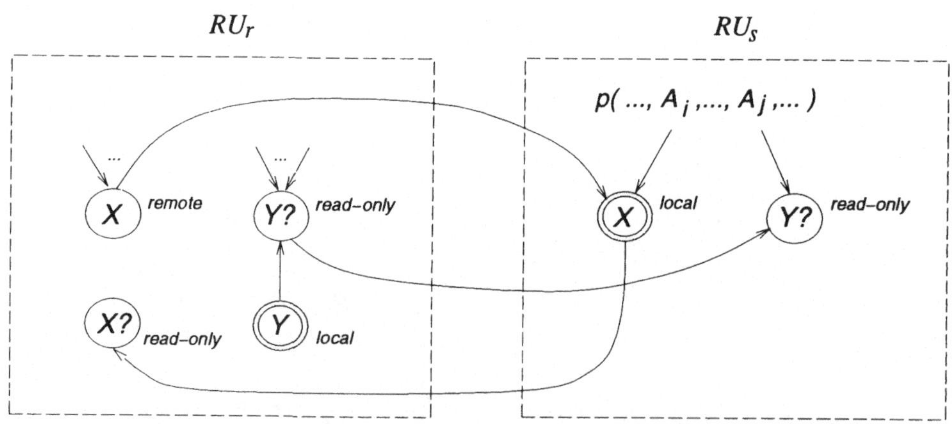

Figure 4.8: Distributed Representation of Processes (Alternative II)

Compared to the first approach, the second one has a considerable advantage. Remote access operations to write-enabled variables as well as the process suspensions they effect do occur far less frequently. Thus, the overhead for communication and synchronization is significantly decreased and the resulting impact on the overall performance becomes reduced by the same degree.

Although it is much more preferable to choose the second alternative for the performance improvements it provides, there still is a particular problem with this solution. In order to convert write-enabled variables belonging to argument structures of process P into $XERs$ pointing from the old variable locations at RU_r to the new variable locations at RU_s, the respective addresses of the new locations have to be known a priori. Unfortunately, these addresses are not available before P has been placed at RU_s. Thereafter, RU_s will be able to communicate the heap indices of the newly created variables back to RU_r. For the duration of time this requires, the pointer values of the corresponding $XERs$ at RU_r remain undefined.

Temporarily undefined XERs, of course, necessitate special handling by the distributed reduction algorithm. On the other hand, the resulting effect on program execution is much like the effect caused by read-only variables. Both kinds of data objects, read-only variables as well as temporarily undefined $XERs$, represent process argument structures which are not yet defined.

4.2.6 Distribution of Process Strucures

When a process structure is copied from one reduction unit to another, the corresponding argument structures belonging to the process structure have to be packed by the sender unit. This is always necessary, as the individual data components forming an argument structure may be arbitrarily distributed over the local storage of a reduction unit. At the same time, the absolute heap addresses referred by pointer objects are translated into relative addresses. The receiving unit therefore can load the process structure in one block by allocating a corresponding number of consecutive heap words.

4.3 The Distributed Reduction Algorithm

In a distributed computation of the parallel FCP machine concurrency is modeled by interleaving sequences of atomic actions. The granularity of these actions most of all determines the behaviour of the distributed reduction algorithm. It should be emphasized that in spite of the fact that the basic operational behaviour of the machine network is asynchronous, there frequently occur situations where certain operations have to be synchronized.

Concurrent process reductions operating on a common set of global variables must not lead to inconsistent variable bindings. Possible conflicts are avoided process reduction operations respectively compute and carry out the resulting variable bindings either all at once or none of it. This is, process reductions have to be performed as *atomic actions.*

Another aspect of global synchronization concerns the realization of the distributed process suspension mechanism. When a process has been successfully reduced, the instantiation of variables within the local environment of that process must have a global effect on related read-only variables. Processes which have been suspended on these variables need to be woke up – independently of the location of the read-only variables.

Global synchronization always necessitates some form of direct interaction between two or more reduction units. The corresponding operations are realized by means of message-passing based communication protocols.

4.3.1   Variable Migration

Whenever a reduction unit encounters a XER in an attempt to access a variable, it cannot access the variable immediately. However, a reduction unit operating as a variable member always may gain access to the variable by changing its *member status* into an *owner status*. Of course, it must not do this on its own, but it may do so under control of the current variable owner.

The operation of exchanging ownership between the variable owner and one of the related variable members is denoted as *variable migration*. In order to initiate a variable migration operation, the respective variable member requests ownership by sending a corresponding message to the current variable owner. The message contains the address of the requested variable at the owner unit, the identifier of the owner unit and the member unit, and the address of the XER at the member unit: *message(VariableRequest, OwnerUnit, MemberUnit, AddressOwner, AddressMember)*.

The applied variable representation scheme ensures that there always exists a path from each variable member to the current variable owner. Variable requests are forwarded accordingly, in case that a path is of length $l > 1$.

When receiving a variable request for variable X from some member unit RU_j, the reaction of the owner unit RU_i depends on the actual state of X. This state is referred as $value(i, X)$. If X is yet an unbound variable, then $value(i, X)$ represents a pointer to possibly existing read-only occurrences of X. That means, $value(i, X)$ either identifies the location of some read-only occurrence X? or $value(i, X) = nil$ in case read-only occurrences do not exist.

An unbound variable X causes RU_i to return $value(i, X)$ to RU_j. Now, RU_j replaces $value(j, X)$ – the pointer value of the XER – through the read-only variable pointer $value(i, X)$. On the other hand, RU_i substitutes $value(i, X)$ by a pointer referring to the current location of X at RU_j. At the same time, RU_i and RU_j respectively change the attributes of X from *local* to *remote* and vice versa. As a consequence, RU_j effectively becomes the new owner of variable X.

Using the example shown in Figure 4.5 (Section 4.2.4), the outcome of a variable migration operation which is initiated by RU_2 for the unbound variable X is illustrated in Figure 4.9.

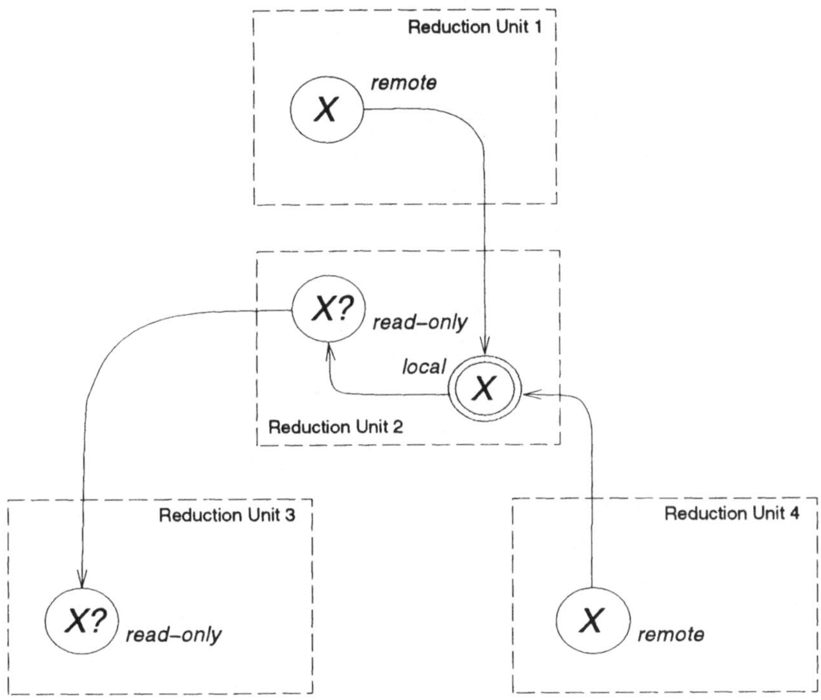

Figure 4.9: Effect of Variable Migration

In the complementary case, if X has already been bound to a non-variable term, RU_i returns a copy of this term in order to replace the remote reference at RU_j. Since the term itself may contain variables, in the structure being copied each of this variables has to be replaced by a corresponding XER. The status of RU_j with respect to the former variable X thereby becomes meaningless.

When considering the timing requirements for the variable migration protocol in detail, it shows that there always results an intermediate state with an invalid representation of X (Figure 4.10). Though this state is visible for any reduction unit interacting with RU_i or RU_j, it cannot lead to an unsecure situation.

The only harm that the invalid representation of X may cause is an almost negligible delay of concurrently performed variable requests referring to X. In a situation where RU_i (RU_1) has already granted the request from RU_j (RU_2) but RU_j has not yet received the corresponding grant message, none of the other member units will be able to obtain the ownership of X. Request messages are simply forwarded until it is recognized that RU_j has become the new variable owner.

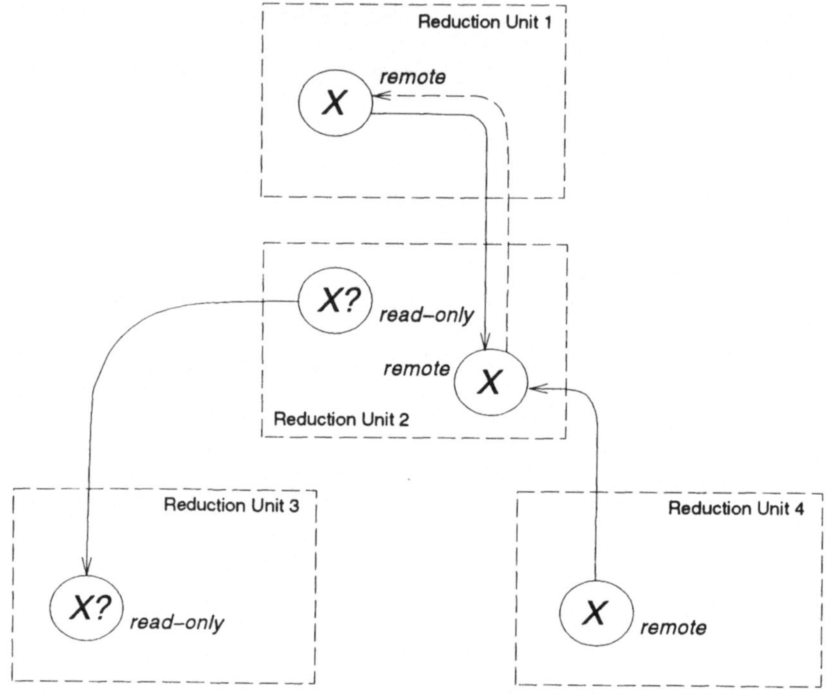

Figure 4.10: Intermediate State in Variable Migration

4.3.2 Synchronizing Reduction Operations

The applied scheme for distributed representation of variables together with status dependent access restrictions as implied by the owner/member relationship guarantees mutual exclusive write access to globally shared variables. Nevertheless, as long as a variable remains unbound, any reduction unit keeping a reference to that variable may also obtain access to the variable by initiating a variable migration operation. The realization of the variable migration protocol is secure in the sense that access conflicts cannot occur while a variable migrates from one reduction unit to another. The resulting features concerning the elementary handling of variables provide the foundation for implementing concurrent reduction operations as atomic actions.

With regard to the general structure of guarded Horn clauses and the typical data manipulation operations in processing it, a *process reduction cycle* always has the following form. It consists of one or more clause try operations; each of which logically divides into two subsequent phases. The first phase handles *head unification* as well as *guard evaluation* prior to commitment; the second phase spawns new subprocesses as defined in the clause's body (see also Chapter 3).

When performing a process reduction cycle, a reduction unit must not be interrupted before one of the following two *stable states* is reached. It either has completed the second phase of a successful clause try operation, or it has restored its old data state by releasing any temporarily computed variable bindings in case that the process becomes suspended. This feature has to be ensured by the implementation of the process reduction mechanism. The implementation details will be considered in Chapter 5.

During the first phase of a reduction cycle variables are affected only within a reduction unit's local environment. If a clause try fails, these effects can easily be undone using a *trail stack*. The trail stack identifies all writable and read-only variables that have been modified during a reduction attempt. For each of this variables there is a trail stack entry containing the type, the pointer value, and the location of the variable: $trail\_stack\,[i] = (VariableType, PointerValue, HeapIndex)$, $\quad 0 \leq i$.

In fact, a reduction unit is able to affect a variable, only, if it is the current variable owner; otherwise, the reduction would be aborted and the process suspended. A successful clause try reaching the commit operator thus requires the reduction unit to be owner of *all* variables being involved in the reduction. At the same time, no other reduction unit can become owner of any of these variables, since the reduction cannot be interrupted. As a result, a reduction globally modifies all involved variables in case it succeeds, while it does not affect any variable neither globally nor locally whenever a clause try fails; hence, it is *atomic*.

4.3.3 Distributed Process Suspension

During a distributed computation there may occur many thousands or even millions of events which necessitate some form of global synchronization. A suitable means to delay local operations according to global time constraints without using busy waiting is the application of a process suspension mechanism. The total number of process suspensions in a particular computation, of course, is a variable depending on various parameters.

First of all, the number of suspensions depends on the specific data-flow synchronization demands that are inherent to the application problem itself. Moreover, there are certain parameters concerning the realization of the parallel machine which also have a considerable impact on the suspension behaviour. Particularly important is the choice of the scheduling policy. This includes both, *local* as well as *global* process scheduling. The later one is substantially affected by the characteristics of the dynamic load balancing algorithm.

Whenever possible, process suspensions should be avoided since they always cause a performance degradation. Primarily, this is achieved by measures that improve local and global scheduling. On the other hand, as it is not reasonable to try to realize an optimal scheduling when using a deterministic scheduling algorithm, there are certain limitations. For that reason, the efficiency of the process suspension mechanism is material to the resulting system performance. Efficiency thereby refers to the costs caused by local suspensions as well as the cost for additional communication which arise when suspending a process on a remote data object.

In any case, process suspensions can only occur in combination with one of two data objects; these are *read-only variables* and *XERs*.

Process Suspension on Read-only Variables

When a variable X becomes instantiated by a reduction unit RU_i, the resulting term T is propagated to all read-only occurrences $X?$ being located at the same or any other reduction unit. The variable representation scheme ensures that there always exists a path via references from the location of X at the owner unit RU_i to each occurrence $X?$ residing at reduction units $RU_{j_1}, RU_{j_2}, ..., RU_{j_k}$ (not necessarily being member units).

In order to identify those processes which have been suspended on a read-only occurrence of the variable X, a possibly empty list of *process suspension notes* is associated with each $X?$. A *suspension note list (SNL)* is attached to a read-only variable $X?$ belonging to some reduction unit RU_j by storing a corresponding list pointer in $value(j, X?)$.

In addition to suspension notes identifying processes within the local environment of RU_j, the *SNL* at $value(j, X?)$ may also contain references to read-only occurrences of X being located at remote units. The former type of suspension notes is called *local suspension note*, while the later type is denoted as *global suspension note*.

Each local suspension note identifies a different process of RU_j which has been suspended on $X?$. Traversal of the *SNL* then allows to wake up these processes in order to schedule them again. By application of a two-stage referencing scheme it is ensured that also in case that a process is suspended on more than one variable at a time, it becomes effectively woke up only once. All multiple references on the same process are invalidated as soon as the first wake-up operation is performed.

A remote suspension note in the *SNL* of $value(j, X?)$ always identifies a set of suspended processes outside the local environment of RU_j. In contrast to a local suspension note, it does not refer to such processes, immediately, but it refers to a remote read-only occurrence of X residing at some reduction unit RU_k $(j \neq k)$.

A remote suspension note is handled by sending a *RemoteInstance* message containing the address of the remote read-only occurrence $X?$ at RU_k *RemoteAddress* together with the resulting term *ResultTerm*:

$$message(RemoteInstance, RemoteUnit, LocalUnit, RemoteAddress, ResultTerm).$$

When this message is received by RU_k, the receiver unit activates processes which are referred in the SNL at value$(k, X?)$. Figure 4.11 gives an example.

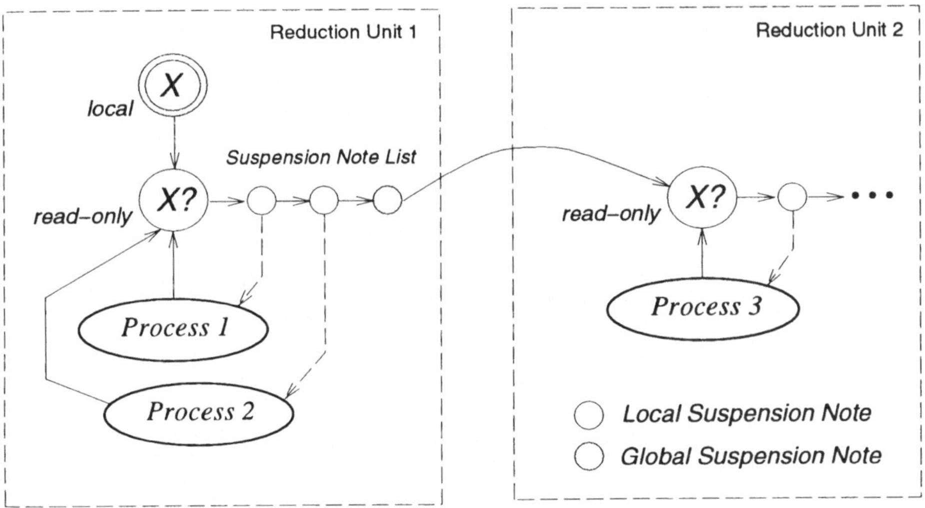

Figure 4.11: Process Suspension on Read-only Variables

Beside *ordinary* local and global suspension notes, a SNL belonging to a read-only variable $X?$ may contain a third type of suspension notes. An instance of this additional type always identifies a *subset* of local or global suspension notes which have subsequently been attached to $X?$. This third type of suspension notes is used in order to efficiently combine the SNLs of two or more read-only variables.

When two variables X and Y are bound with one another, the second one becomes a pointer to the first one (or vice versa). If both variables refer to respective read-only variables $X?$ and $Y?$, the read-only variables are handled the same way. This means that the read-only variable $Y?$ is converted into a pointer referring to $X?$.

Now, assuming that each of the read-only variables holds a non-empty SNL, then the suspension notes of both lists have to be merged into a common list belonging to the remaining read-only variable $X?$. This is carried out in one step by adding an additional suspension note to the top of the already existing SNL of $X?$ (Fig.4.12).

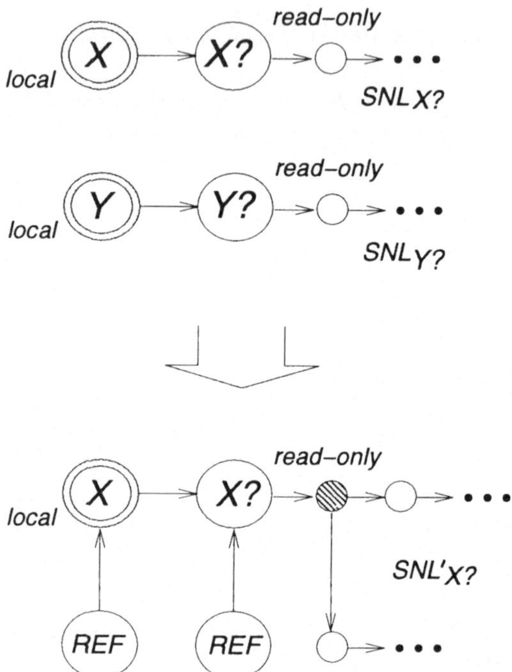

Figure 4.12: Unification of Variables

Process Suspension on XERs

Just the same way as suspension notes are used in order to suspend processes on
read-only variables, they can also be used in combination with *XERs*. A variable
request needs always to be performed for the first process suspending on a *XER*,
only. Thereafter, the *pointer field* of the *XER* can be reused to store a pointer
to a *SNL* keeping the process which has initiated the variable request. Additional
suspension notes can be added in case that other processes try to access the same
XER. When a *XER* is changed into a local variable or it is replaced by a resulting
nonvariable term, all processes that have meanwhile been suspended on this *XER* are
woke up.

An example configuration is illustrated in Figure 4.13 and Figure 4.14. The configuration shown in Figure 4.14 is obtained if first *Process 1* and thereafter *Process 2* attempt to access the *XER* referring to variable X at *RU 2*.

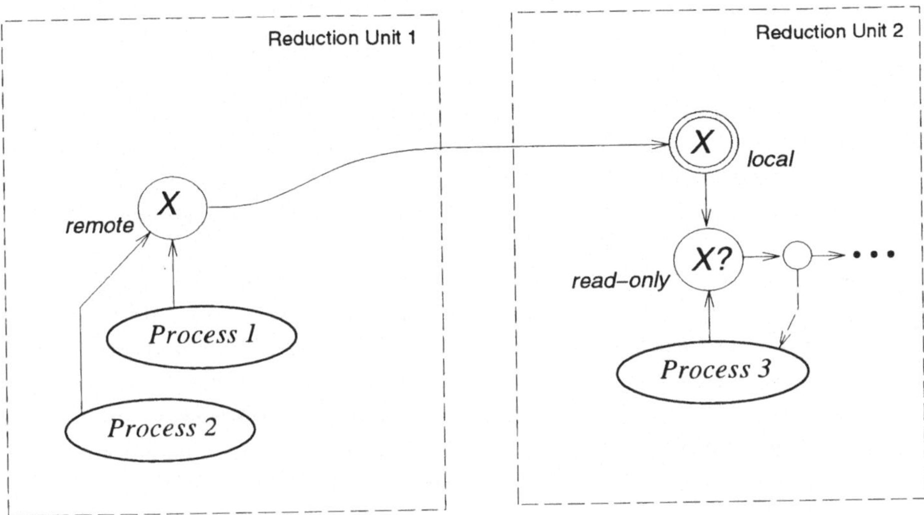

Figure 4.13: Process Suspension on XERs: Initial Configuration

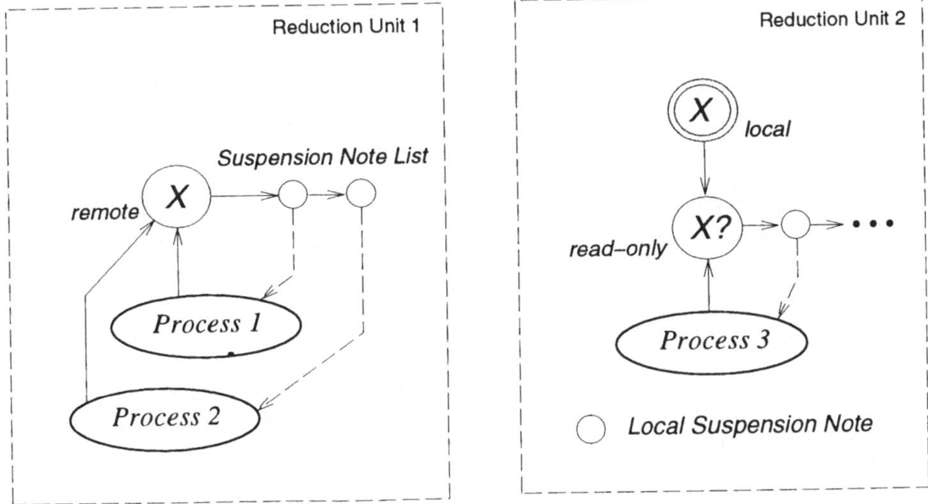

Figure 4.14: Process Suspension on XERs: Resulting Configuration

4.3.4   Observations on Complexity

Compared to a purely sequential reduction algorithm, the additional complexity of the distributed reduction algorithm is directly related to the message transfer rate and the complexity of the messages being transferred. Among these messages it can be distinguished between messages related to *distributed control and synchronization* and messages concerned with *global structure copying*. The later type of messages includes *load transfer* messages, which are used to distribute process structures, as well as *remote instance* messages, which are used to copy individual argument structures.

Compared to the complexity of messages containing process or argument structures, the complexity of control and synchronization messages is meaningless (usually they consist of a message type identifier and a few global address pointers). However, the relative frequency of control and synchronization messages has a crucial impact on the overall performance behaviour, since they contribute the largest amount among all the messages. An important role in the attempt to reduce this amount as far as possible has the choice of the distributed variable representation scheme (cf. Section 4.2.4) as well as the realization of the distributed process suspension mechanism.

With regard to events causing a message transfer operation which is concerned with distributed control or synchronization [3], the proposed solution is characterized by the following behaviour:

- In respond to each load transfer operation distributing any number of processes, a single *variable location message* is produced. It contains a list of variable addresses indicating the new locations for the variables that have been moved to the receiver unit.

- An initial request on a *XER* which has not yet been requested produces a *variable request message*. Subsequent requests on the same *XER*, however, do not cause further messages.

- Each time a local read-only instance of a *XER* needs to be created and this *XER* has not yet been requested, a *read-only location message* is sent to the variable owner referred by the *XER*. This way, the newly created read-only variable is included into the corresponding reference chain identifying the related read-only variables.

---

[3]This includes neither messages belonging to the distributed termination detection procedure (in fact, these are negligible as will be seen in Section 4.4) nor messages belonging to the dynamic load balancing procedure.

It should be noticed that the local suspension of a process on one or more read-only variables does *never* cause any message transfer operation. The assignment of values to read-only variables is carried out automatically as soon as their write-enabled counterparts become instantiated. At the same time, the suspended processes become rescheduled again. Explicit *read request messages* hence are superfluous. This observation is particularly important, as situation in which an attempt is made to read a remote term before this terms is available occur very frequently.

Compared to other approaches using explicit read request messages to be recorded by means of suspension notes which have to be attached to the physical variable at the owner unit, e.g. as proposed in ([Taylor87],[Taylor89]), this implicit evaluation scheme seems to be superior, especially for large-scale networks. According to [Taylor87], an investigation of the communication behaviour of a number of example programs showed that about 90 percent out of the total number of messages transmitted for both, reading and writing of remote terms, are concerned with reading and transferring values.

4.4 Distributed Termination Detection

The current version of the parallel FCP machine realizes distributed termination detection by means of a 2-phase termination detection algorithm. This algorithm corresponds to a straightforward extension of an algorithm which was proposed by Dijkstra et al. in [Dijkstra83]. The extension allows to detect both, the *successful termination* of a computation as well as a *computation deadlock* corresponding to a situation in which all remaining processes have been suspended.

Global termination of a distributed computation, when carried out on a network consisting of n reduction units $RU_0, ..., RU_{n-1}$ $(n > 1)$, is observed by the host unit. In the protocol that realizes the termination detection algorithm, the host unit therefore takes the role of a central supervisor. In particular, it initiates and finishes the necessary termination detection cycles.

Immediately, after the host unit has started a computation by sending an initial process to one of the reduction units, it initiates the first termination detection cycle by generating a *termination detection token*. This token carries two kinds of information: a *token colour* and a *suspension count*. An initial token has the colour "white" and its suspension count is set to "0".

The termination detection token is used to represent information about the *global* computation state of the network. The corresponding *local* computation states of the reduction units are represented in a similar way. To each of the reduction units a colour is assigned, as well. In addition, the reduction units are supplied with suspension counters which are incremented and decremented according to the number of locally suspended processes.

Initially, all reduction units have the colour "white". The initial colour of a reduction unit RU_i is changed to "black" when RU_i produces a message addressing some unit RU_j which might have the effect to induce new work at RU_j or any other reduction unit RU_k $(0 \leq i, j, k \leq n - 1)$ (i.e. a work grant message, or any message that could wake up suspended processes).

Assuming a predefined ordering of the reduction units, e.g. $< RU_0, ..., RU_{n-1} >$, an initial termination detection token is propagated through the network such that it is passed from the host unit to RU_0, from RU_{i-1} to RU_i $(1 \leq i \leq n - 1)$, and from RU_{n-1} back to the host unit. When a reduction unit RU_i receives the token, the unit keeps the token local as long as it remains busy.

As soon as RU_i becomes idle, it acts as follows. If the colour of RU_i is "white", it leaves the token colour unchanged but increments the suspension count of the token according to the current number of locally suspended processes. Otherwise, if the colour of RU_i is "black", it sets the token colour to "black" and its own colour to "white". Thereafter, RU_i delivers the resulting token to the subsequent unit.

By means of the colour of a received token, the host unit recognizes global termination. More precisely, in order to detect global termination of the distributed computation, it always requires to run at least *two* termination detection cycles.

When the host unit receives a *white* token and the previously received token has also been a *white* one, it knows that all reduction units are idle and cannot become busy again. Depending on the suspension count s carried by the token, it either signals successful termination $(s = 0)$ or it signals that the whole computation has been suspended $(s > 0)$.

When the host unit receives a *black* token or the previously received token was *black*, it starts a new termination detection cycle by generating an initial token as explained above.

It is important to realize that the extension of the original algorithm does not affect its correctness. The crucial issue is that the added suspension count does not have any impact on the termination behaviour of the algorithm itself.

The particular advantage of the described termination detection algorithm, beside simplicity, is the relatively small overhead to be paid. In fact, the performance degradation caused by this algorithm is almost negligible, since a termination detection token cannot pass a reduction unit as long as the unit remains busy. This way, the relative frequency of termination detection cycles that are performed during a computation is quite small when there is a large number of operating reduction units.

Unfortunately, the algorithm also has a disadvantage. It does not allow to handle more than one computation at a time.

4.5 Multiprogramming Facilities

Except for the distributed termination detection algorithm, the parallelization concepts embodied in the distributed execution model of the parallel FCP machine do not make any use of the fact that all processes belong to the same computation. Especially for a reduction unit, it does not make any difference whether its local subresolvent contains processes of a single computation or a number of concurrently executed computations.

By application of a more sophisticated termination detection procedure, which allows to handle more than one distributed computations independently of each other, multiprogramming would also be possible; even though, appropriate procedures for distributed termination detection can add a considerable amount of overhead. Examples are algorithms using process counters [Ichiyoshi87] or weights [Rokusawa88] as well as the *short-circuit* technique ([Weinbaum87],[Shapiro89]).

4.6 Deadlock and Livelock Prevention

The parallel FCP machine provides a typical example of a system in which a number of concurrently operating processes require access to *globally shared system resources*. Within the distributed execution model of the parallel FCP machine, the logical variables may be considered as globally shared system resources. From competing attempts to bind these variables there may arise various kinds of access conflicts.

Frequently addressed problems, when dealing with such a system configuration, are the detection and prevention of deadlocks and livelocks. Essentially, these problems are concerned with system resource management and the behaviour of protocols which realize the corresponding mechanisms. The possibility of system deadlocks and livelocks can be eliminated in one of two basic ways. First, the resource management facilities have to be realized in such a way that situations which might lead to deadlocks or livelocks are excluded. Second, provided that deadlocks and livelocks can always be detected without consuming too much of computation time, it is often possible to resolve them immediately.

4.6.1 Prevention of Livelocks

A livelock of the parallel FCP machine would correspond to a situation in which two or more reduction units simultaneously attempt to obtain ownership of the same set of global variables. In order to enable the reduction of a certain process, each of the competing reduction units needs to be owner of the *complete* set of variables. A situation like this might produce a sequence of concurrently performed variable migration operations without reaching a stable state which allows one of the reduction units to successfully reduce its process. The interactions between the competing reduction units correspond to a behaviour that may be characterized as *circular variable stealing*. In the worst case, this could result in an infinite loop.

In order to obtain a livelock, the involved reduction units must behave extremely symmetric with respect to scheduling and communication. Indeed it is not very likely that such situations do appear frequently. However, it might be possible. Moreover, there is another aspect to be considered in connection with livelocks. Beside real livelocks, there also may occur situations in which a possible livelock becomes resolved but the time this requires is not acceptable.

An appropriate means to prevent livelocks is the application of a *variable locking mechanism* [Taylor87]. In an attempt to collect a set of variables, all belonging to the same process, it would be helpful if a reduction unit could *lock* variables that are already local against migration. The locking can be released as soon as the process has been reduced. Though variable locking would eliminate the possibility of livelocks, it also introduces a new problem as it offers the possibility of deadlocks.

4.6.2   Prevention of Deadlocks

A system deadlock can only occur in combination with four conditions that must hold simultaneously ([Coffman71],[Tanenbaum87]): *mutual exclusion, hold and wait, no preemption*, and *circular wait*. In terms of the distributed execution model of the parallel FCP machine, these general conditions have the following special meaning:

(1) *Mutual Exclusion Condition:* Each variable is either currently assigned to exactly one reduction unit or is available.

(2) *Hold and Wait Condition:* Reduction units currently holding variables granted earlier can request new variables.

(3) *No Preemption Condition:* Variables previously granted cannot be forcibly taken away from a reduction unit. They must be explicitly released by the reduction unit holding them.

(4) *Circular Wait Condition:* There must be a circular chain of two or more reduction units, each of which is waiting for a variable held by the next member of the chain.

A variable locking mechanism as described above, obviously implies the risk of system deadlocks. In particular, it would allow that reduction units maintain variable locks unless they have locked a complete set of variables which is required in a certain process reduction. This form of *partial allocation* has the effect that processes remain suspended forever, in case that a reduction unit does not manage to complete the required set of variables.

In order to avoid deadlocks, the locking mechanism has to be realized in such a way that at least one of the conditions (1) to (4) does not hold. This is easily achieved, when the reduction units are supplied with different *priorities*. Regardless whether a variable has been locked or not, a reduction unit RU_i would always be successful when requesting this variable from a reduction unit RU_j if the priority of RU_i is higher than that of RU_j . This way, a deadlock cannot occur, as the reduction unit with the relatively highest priority among a set of competing reduction units will always be able to get all required variables.

Unfortunately, priorities alone do not eliminate all sorts of deadlocks. There still remain problems that may occur in conjunction with read-only variables. For instance, the reduction unit RU_i with the highest priority might have locked some variable X in an attempt to reduce a process $p(..., X, Y?, ...)$. Now, this process becomes suspended on the read-only variable $Y?$. The corresponding write-enabled variable Y together with a process $q(..., X, Y, ...)$ is assumed to be located on some other reduction unit RU_j. At the same time, RU_j cannot reduce the process $q(..., X, Y, ...)$, which would have the effect that variables X and Y both become bound because this would require the variable X to be local on RU_j. In fact, this means that the described configuration represents a deadlock situation, as well.

In principle, also problems of this kind can be solved. For example, one could use rotating priorities instead of fixed ones [Taylor89]. This way, the problem disappears. Priority rotation would also be helpful to attack *starvation* which is a fairness problem occurring in conjunction with priorities. Starvation refers to the fact that reduction units with low priorities will not have the chance to obtain variables that are frequently locked by reduction units with higher priorities. Nevertheless, rotating priorities can eliminate the problem of starvation, only, if the given timing constraints are also taken into consideration. That is, a reduction unit should have a high priority at least for the period of time it requires to successfully request and lock a desired set of variables.

So far, it seems that all the difficulties which arise when using a variable locking mechanism in order to prevent livelocks can be circumvented. On the other hand, the overhead to be paid for deadlock prevention increases more and more. Hence, a heterogeneous approach based on a combination of *livelock detection* and *deadlock prevention* is proposed.

4.6.3   Detection of Livelocks

A livelock can be detected by observing the number of variable requests that are performed for each individual XER. In fact, this requires an additional counter value to be stored in conjunction with the XER. This can either be done directly by attaching the counter value to the XER or by using a separate table containing the corresponding counter values. Each time a new variable request is started, the counter entry belonging to the related XER is updated accordingly. When the XER becomes eventually replaced by an instance of the corresponding variable, the counter entry can be deleted.

In a livelock situation there always exist $XERs$ for which an *infinite* sequence of variable requests is produced. In a computation which is free of livelocks and deadlocks, however, the maximum number of variable requests involving the same XER is relatively small. Though an upper bound for this number cannot be specified in general, as it depends on the particular application, it is useful to assume some value r as the maximum number of *acceptable* variable requests for the same XER. In most applications, especially those which do not allow multiple writers, there is at most a *single* request for each XER (This value was obtained when investigating a number of example programs on large networks).

Variable request counters when used in combination with a predefined bound r provide a suitable means to detect any situation which might result in a livelock. It is important to observe that the ratio of the number of indicated livelocks, which also includes pseudo-livelocks, to the number of real livelocks can be adjusted by increasing or decreasing the value of r. Choosing a very large value for r would mean that a large amount of computation time is wasted in case a real livelock occurs. Hence, the bound r should be considered as a parameter that is to be determined depending on the particular class of application programs.

4.6.4   A Combined Approach

As real livelocks do not occur very frequently and it is easily possible to detect them, the original problem of livelock prevention can be treated much more efficiently by combining *livelock detection* with *deadlock prevention*. The distributed reduction algorithm is therefore supplied with two distinct *operating modes.*

When operating in the first mode, which is the default one, it avoids variable locking but ensures livelocks to be detected using variable request counters. If a possible situation for a livelock occurs, the algorithm immediately switches to the second operating mode.

Now, reduction units are allowed to lock variables against migration. While the algorithm operates in the locking mode, deadlocks are avoided by assigning rotating priorities to the reduction units. As a consequence, the livelock becomes eliminated. As soon as at least one of the variables which have been identified in connection with the livelock becomes bound, the distributed reduction algorithm again switches back to the default mode.

The advantage of the described algorithm rests upon the fact, that the overhead required for livelock detection is relatively small compared with the performance degradation caused by an deadlock prevention protocol [Taylor89]. Since both events, livelocks as well as deadlocks, do not appear very frequently, this approach provides a general optimization as long as some reasonable bound r is selected.

The proposed algorithm, in any case, is as good as a simple algorithm based on variable locking as the trace of variable requests can be switched off while operating in the locking mode.

4.7 Dynamic Work Load Balancing

In the distributed computation model described, the basic unit of parallelism is the FCP process. When a computation is carried out on a network of asynchronously operating reduction units, the work load is balanced as FCP processes migrate between reduction units. An individual FCP process thereby defines the smallest unit of load that can be transferred between two reduction units. In general, a single load transfer operation may effect any number of process migrations n $(n \geq 1)$.

The number of migrating processes involved in a load transfer operation, however, does not provide a suitable measure for the effective *quantity of load* that is transferred from one reduction unit to another. At any given time during a computation, the various processes building up the process network may represent quite different quantities of work load. As each individual process actually represents a particular subtree within the global computation tree, the execution of a process may correspond to an arbitrary complex subcomputation (see Figure 4.15).

With regard to the costs caused by dynamic load distribution, it is always desirable to transfer a certain quantity of load from one reduction unit to another by means of a minimum number of migrating processes. In particular, this would reduce the costs for packing and unpacking process structures as well as the related communication costs. For that reason, it is important to have an estimation of the quantity of load associated with each individual process.

Even more relevant as the above costs may be the costs that must be paid for additional communication and synchronization, whenever the distribution of processes is in conflict with the locality of computation and communication. In fact, this relationship strongly depends on the particular data interdependencies of the application problem.

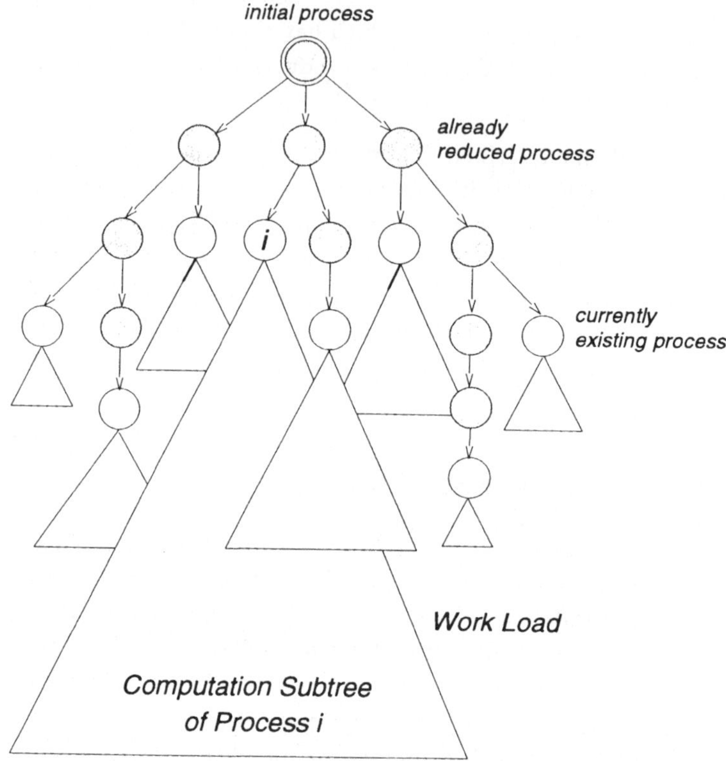

Figure 4.15: Load Characteristic of an FCP Process Network

Obviously, there are at least two fundamental aspects to be considered in conjunction with dynamic load balancing; each of which is of similar importance for the resulting efficiency. First, this is the applied *load balancing policy* determining the relative frequency of load balancing operations, the processors between the load is transferred, and the quantity of load to be transferred respectively. The second aspect refers to the applied *process selection policy* determining which processes among the processes contained in a local resolvent are most suitable for migration. In other words, the load balancing policy is more concerned with the *quantitative* aspects of dynamic load balancing, while the process selection policy is more concerned with the corresponding *qualitative* aspects.

4.7.1 The Process Selection Policy

An appropriate measure to estimate the work load associated with a certain process would be the complexity of the subcomputation required to execute this process. Provided that the whole subcomputation runs locally at the same processor, the complexity is determined by the number of necessary reduction operations in combination with the complexity of each of these reductions.

Regarding a given program $\mathcal{P}$, one could assume appropriate *complexity constants* assigned to each program procedure $C^{p/k} = \{C_1^{p/k}, ..., C_l^{p/k}\} \subseteq \mathcal{P}$ $(l > 0)$. The constant $c_{p/k}$ of a procedure $C^{p/k}$ should indicate the *mean complexity* of a reduction operation performed by means of a clause $C_i^{p/k} \in C^{p/k}$ $(1 \le i \le l)$. The value of $c_{p/k}$ essentially depends on the complexity of the operations necessary for *clause selection* and *clause evaluation* (cf. Section 3.1).

At the same time, the number of reduction operations to be performed in the subcomputation represented by each individual process often depends on the input size. Even if the computation is finite, the input size represents a figure which usually cannot be determined at run-time. In fact, this means that the actual complexity of such a subcomputation is not known before the whole computation has been terminated, successfully.

A Heuristic Approach. Parallelization which is performed directly at the layer of individual FCP processes provides an extreme. As the number of processes usually is very large and an individual process often represents a relatively small unit of computation, it seems to be reasonable to decrease the granularity of parallelization. Processes should be combined in order to obtain a smaller number of more complex units. A corresponding approach for classifying processes at run-time using a heuristic is based on *hierarchical process clustering* [Glaesser91a]. The central aspect under which processes are grouped into clusters is to retain a relationship between those processes most likely to communicate with each other.

Hierarchical process clustering is realized by organizing the local subresolvents as a hierarchy of k different layers $l_0, l_1, ..., l_{k-1}$, where k is a variable. Each layer l_i $(0 \le i \le k - 1)$ consists of a possibly empty set of process clusters $C_{i,0}, ..., C_{i,r}$. When a process belonging to a cluster $C_{i,j}$ becomes reduced, the resulting subprocesses are included into a cluster $C_{i+1,l}$, which is attached to $C_{i,j}$ using a corresponding reference.

This way, a local subresolvent is partitioned into a number of hierarchically embedded process clusters. Depending on the choice of k, the depth up to which processes are clustered can be limited. Each cluster respectively identifies a set of processes belonging to the same subcomputation. The particular advantage of the described hierarchical organization scheme is that it offers almost direct access to these processes.

When performing dynamic load balancing, it is desirable to partition the local resolvent according to the ordering implied by the computation tree. The resulting communication and synchronization costs can be reduced if the set of migrating processes forms a closed subcomputation. This is achieved, when determining the migrating processes by selection of a suitable subcluster.

Unfortunately, there still remains an open problem concerning the handling of suspended processes. The overhead caused by the proposed scheme is reasonable small and can be properly adjusted by the choice of k, as long as suspended processes are ignored. On the other hand, the whole procedure becomes more and more ineffective, when the number of process suspensions increases.

Experimental results have shown that hierarchical process clustering which ignores suspended processes is useful only to improve the initial load distribution at the beginning of a computation, but it should be switched off thereafter unless a way is found allowing to handle suspended processes, efficiently.

4.7.2　The Load Balancing Policy

The primary goal of dynamic load balancing is efficient utilization of the computational power offered by a given number of available processors. Standard approaches usually try to achieve this goal by minimizing the idle times of the processors employed. Apart from increasing the computation speed, a secondary goal of dynamic load balancing may be the uniformly utilization of local storage.

In contrast to dynamic load balancing approaches which transfer basic parts of the control over load balancing activities into the program ([Taylor89],[Takeda90]), the approaches considered here realize system controlled load balancing.

In combination with the distributed reduction algorithm of the parallel FCP machine divers load balancing algorithms have been implemented and investigated. The various strategies realized by the applied algorithms more or less provide a representative choice among known load balancing models. More secifically, about 10 strategies based on the following models have been investigated:

o *nearest-neighbour models*

o *gradient models* [Lueling91]

o *load-oriented models* [Lueling92]

o *adaptive models* with centralized control [Xu90]

(A central supervisor frequently adapts certain control parameters affecting the global load balancing behaviour to changing load situations. The actual load balancing activities, however, remain completely decentralized.)

An extensive comparison of the referred strategies was based on experimental results obtained from running a number of smaller application examples on large-scale Transputer networks with different topologies. The applied examples include standard test programs such as *N Queens* [Okumura87], *Towers of Hanoi* [Houri87], *Quicksort* [Shapiro86] etc.

The outcome of the comparison was that none of the tested strategies behaves superior to all the others. Especially on very large networks ($n \geq 256$ Processors), a relatively simple, load-oriented strategie based on local-search and local-distribution [Lueling92] for certain application examples showed to be as good as and sometimes even better than the much more sophisticated strategies. The results of this comparison will be summerized in [Kaercher92].

4.8 Distributed Garbage Collection

FCP as well as many other symbolic programming languages relies on dynamic storage allocation, which means that dynamically created language objects are not explicitly destroyed at run-time. In order be able to use the parallel FCP machine for real applications, it must be supplied with an efficient mechanism for system-wide garbage collection.

The problem of distributed garbage collection is not further investigated here as the target architecture for the prototype implementation offers sufficient storage to run application examples of a considerable size (cf. Section 5.1). Moreover, there already exist various garbage collection algorithms for distributed-memory machines. For example, an asynchronous garbage collector for message-passing multiprocessor systems, as described by Foster in [Foster89], might provide a suitable solution.

4.9 Related Work

In recent years, a considerable number of concurrent logic programming languages has been proposed and there already exist implementations for many of these languages. Meanwhile, main research activities primarily concentrate on the investigation of *flat languages*, which seem to be particular suitable for efficient implementation [Shapiro89]. Nevertheless, most of the implementations of these languages are uniprocessor implementations. Up to now, there are only a few distributed implementations running either on non-shared-memory multiprocessor or on multicomputer systems ([Ichiyoshi87],[Foster88],[Taylor89],[Foster90],[Nakajima92]).

In some sense the above language implementations, more or less, are all related to the one presented here. On the other hand, even when restricting on flat languages, there are significant semantic differences which may have a strong impact on the language implementations. Especially, the various synchronization techniques used to specify data-flow constraints lead to rather different parallel execution models. Languages that do not allow unification of variables prior to commitment to a clause, e.g. Parlog [Foster88], Strand [Foster90], or KL1 [Nakajima92], completely avoid situations in which processes compete for variables; hence, problems as livelocks and deadlocks cannot occur and even variable migration operations become superfluous.

Other FCP Implementations. Closely related to the described work is an implementation of FCP at the Weizmann Institute of Science. An instruction set for a sequential FCP machine based on a modified *Warren Abstract Machine* [Warren83] together with an FCP compiler written in FCP was proposed in [Houri87]. An initial parallel interpreter for FCP running on an iPSC Hypercube is described in [Taylor87]. Various compilation techniques for an optimizing FCP compiler as well as an improved version of the initial parallel execution algorithm are presented in [Taylor89]. Despite of the fact that both approaches attempt to implement the same language, there are a number of substantial differences with respect to the way this is realized. They are discussed below.

Sequential Implementations. First of all, the sequential machine [Hannesen91] employed in our own implementation is based on an instruction set distinct from the one used in [Houri87]. An FCP compiler for this machine has been developed by means of the *Eli* compiler construction system [Gray92].

On the other hand, this point is actually not material to the subsequently described implementation of the parallel FCP machine, which is constructed in such a way that the sequential machine components can easily be replaced by any other realization of a sequential FCP machine (see also Chapter 5). Of considerable importance, however, is the design of the underlying distributed run-time system, especially the algorithms concerned with distributed unification, distributed control, and dynamic load balancing.

Distributed Variable Representation. The fundamental data structure for the distributed reduction algorithm is the representation scheme for globally shared logical variables. In fact, this data structure represents something like the *backbone* of the whole parallel machine with regard to efficiency and flexibility. In particular, it determines the message transfer rate, which is an important factor when operating on large-scale networks.

Conceptually, both representation schemes, the one suggested in [Taylor87] as well as the one applied here, realize some kind of restricted global address space by adopting well-known techniques also used to solve the *data coherence problem* in traditional multiprocessor systems with multiple caches [van de Goor89]. Nevertheless, they do this in a substantially different manner. The obtained solutions lead to rather different *synchronization schemes* with respect to both, the way distributed

process suspension is realized as well as the resulting communication costs (see also
Section 4.3.4). The difference between these schemes can be characterized by analogy to the different control mechanisms associated with event-driven and time-driven
simulation.[4]

Distributed Control. There are two basic system tasks concerned with distributed control: global termination detection and prevention of livelocks. While
global termination detection in both implementations relies on the same distributed
algorithm [Dijkstra83], the problem of livelock prevention is treated in different ways.

In [Taylor87] a variable locking mechanism based on priorities assigned to processors is suggested. In order to resolve possible deadlocks that may result from
circular variable locks, a deadlock detection algorithm is employed. However, even in
the improved version, the deadlock detection algorithms cannot avoid the problem of
starvation [Taylor89].

In contrast to this relatively complex and costly approach, a livelock prevention
algorithm using a combined strategy based on livelock detection and deadlock prevention, as presented in Section 4.6, seems to be much more efficient; especially, as
livelocks can only occur in multiple-writer programs and even there they are very
unlikely.

Dynamic Load Balancing. The problem of dynamic load balancing, in principle, can be handled in one of two basic ways. Control over dynamic load balancing
either remains under the responsibility of the programmer or it is delivered to the
distributed run-time system. Although both solutions have their own advantages
as well as disadvantages, it seems to be reasonable to have the optional possibility
to leave as much control as desired with the system. This necessitates to treat the
dynamic load balancing component as an integral part of the system design, which
is in contrast to the approaches followed in [Taylor89] or [Foster90].

---

[4]This analogy was contributed by Prof. F. J. Rammig.

Chapter 5

Implementing FCP on Large Transputer Networks

5.1 Parallel Machine Architecture

The basic concepts of how to implement the concurrent logic programming language FCP on a non-shared-memory multiprocessor architecture have been introduced in Chapter 4. The concepts as described so far are applicable to a wide class of parallel and distributed computer systems. In particular, they are suitable for all those systems which share the same elementary system attributes as defined by the abstract system architecture in Section 4.1.

A concrete realization of the parallel FCP machine, of course, requires a more detailed specification of the *micro-architecture* of its basic building blocks and the way they interact with each other. A specification of the micro-architecture thus refers to the construction of the uniformly realized reduction units and the host unit.

The remainder of Section 5.1 outlines some more basic issues which are relevant for implementing the parallel FCP machine on large Transputer networks. The subsequent sections present the details.

Modularization. When designing the micro-architecture of the parallel machine, special efforts have been taken to achieve a strict and clear modularization. As far as the reduction units are concerned, the chosen modularization reflects a functional separation according to *communication tasks* and *operation tasks*. The operation tasks are solely devoted to the execution of a reduction unit's sequential FCP machine component. In contrast to the operation tasks, the various communication tasks refer to the underlying distributed run-time system. In particular, they also include general *housekeeping tasks* such as dynamic work load balancing or distributed termination detection.

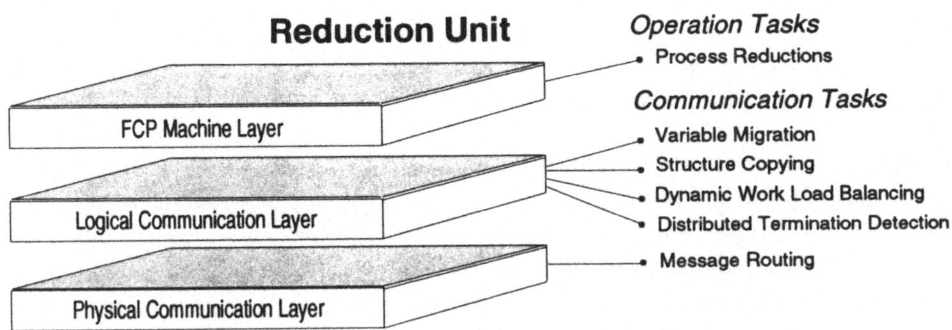

Figure 5.1: Basic Operating Layers of a Reduction Unit

In addition to a functional separation into communication and operation tasks, the modularization also reflects a distinction of three basic layers which are identified with the submodules and their associated functions (see Figure 5.1).

Dynamic Configuration. The parallel FCP machine, as already mentioned, is designed to automatically adapt to arbitrary network topologies. An initially unknown network is explored under control of the host unit, setting up a uniquely identified reduction unit at each allocated processor.

Depending on the number of available processors, the resulting machine configuration on which the FCP program is to be executed may have just a single reduction unit operating completely sequential, or it may have any number n $(n > 1)$ of asynchronously operating reduction units. Nevertheless, it is not necessary to know the number of executing reduction units in advance, i.e. before a computation is started. For that reason, the described realization of the parallel FCP machine on Transputer networks should be considered more as a *generic algorithm* generating suitable machine instances rather than a static algorithm for a particular parallel machine architecture.

A canonical mapping of an instance of the parallel machine consisting of $n + 1$ processing elements – the *host unit* together with n *reduction units* – onto an equally sized Transputer network means that a special processor is reserved for the host unit, while the remaining n processors operate as reduction units.

Each of the uniformly constructed reduction units may be directly connected with up to 4 neighbour units via external links. In fact, the maximum number of possible links is not determined by the reduction unit architecture but is a restriction placed by the Transputer system. The reduction unit architecture would easily scale to any number of links that are made available by the underlying hardware system.

The special processor which runs the host unit must be supplied with the necessary input/output facilities, in order to load the application program as well as the input data and to output the result. In contrast to the host unit, the reduction units do not require such additional facilities.

Implementation. The parallel machine is realized as part of a special programming environment providing a number of useful tools for distributed debugging and run-time evaluation. The whole system has been implemented on Transputer networks using the parallel programming language *Par.C* [Parsec89]. The hardware platform is a fully reconfigurable Transputer system architecture with up to 320 T800 Transputer nodes [Funke92] (see also Chapter 6).

The language Par.C was chosen, as it combines the usual high-level language constructs of standard C with the well-known parallel language constructs of the original Transputer language Occam [May85]. Communication and synchronization between parallel processes is expressed and controlled by means of the data type *channel*, the *select-alt*-construct, and the *par*-construct almost the same way as in Occam. Nevertheless, Par.C also offers a number of additional options which go beyond the parallel execution model of CSP [Hoare78].

5.2   Reduction Unit Architecture

A reduction unit logically divides into two functional subunits, a communication unit and a sequentially operating FCP machine. The sequentially operating FCP machine is represented by a single module called *Reducer*. The communication subunit, however, is further split into a *Router* module and a *Distributor* module. The reduction unit architecture is illustrated in Figure 5.2

The above modules are implemented by a number of parallel processes running on the same Transputer node. While the Reducer and the Distributer are implemented by a single process, respectively, the Router is implemented by a collection of parallel processes. More precisely, instead of *parallel* one should better use the term *pseudo-parallel*, since an individual Transputer can always execute one process at a time.

Processes belonging to the same reduction unit communicate with each other in one of to ways. First, they communicate through predefined *communication channels*. Second, they communicate by sharing *global data structures*. Communication primarily concerned with synchronization is modeled via communication channels. This way, the control structures become visible. On the other hand, when communication is concerned with the exchange of data rather than with synchronization, it is sometimes convenient to use globally shared data structures instead of communication channels.

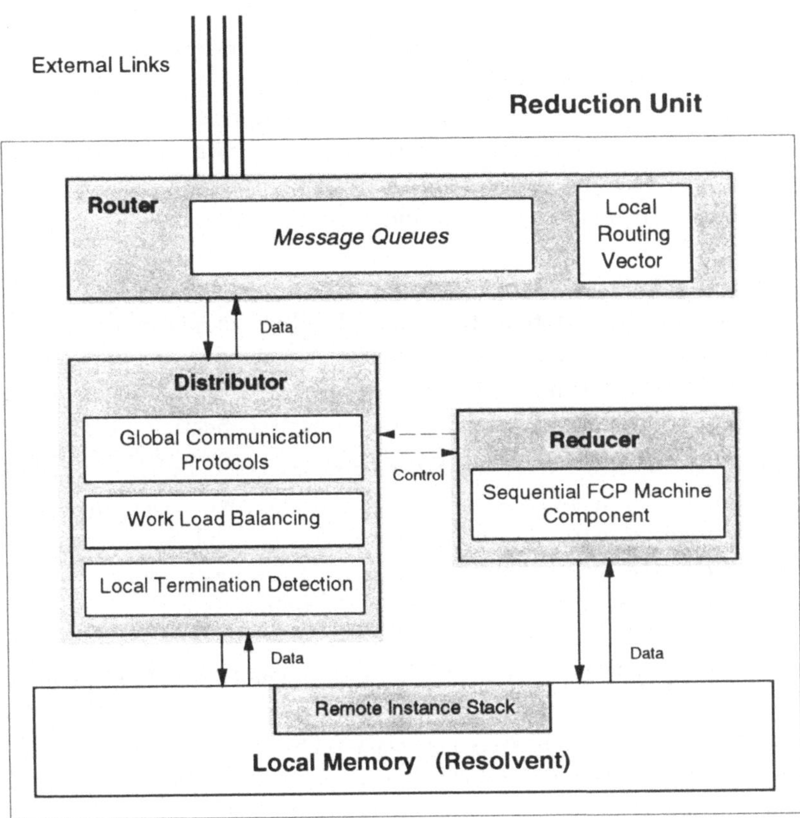

Figure 5.2: Reduction Unit Architecture

The functional separation of the communication unit into a Distributor and a Router refers to the distinction of communication depending on the layer it is carried out (cf. Figure 5.1). The Distributor deals with communication at the *logical layer*, where interaction between reduction units are expressed by means of communication macros. This offers the possibility of specifying high-level communication protocols without bothering about low-level communication requirements. In particular, this allows to assume an asynchronous communication model for global system communication.

In contrast to the Distributor, the Router is concerned with communication at the *physical layer*. Beside translating the communication macros for the Distributor into corresponding low-level communication-protocols, it handles the transformation of asynchronous into synchronous communication, and vice versa.

5.2.1   The Reducer Subunit

Operating as a local reduction engine, the Reducer represents the actual core component of a reduction unit. As the Reducer does not have any own facilities for global communication, it needs to invoke the Distributor whenever interactions with other reduction units become necessary. There are exactly two situations where such interactions occur. The first one is concerned with *variable migration operations* and the second one with *remote instantiations.*

Remote Instantiations. In order to instantiate a read-only variable outside the local environment of a reduction unit, the Reducer places the necessary information on a so-called *Remote Instance Stack.* A new entry for the *Remote Instance Stack* is generated each time the Reducer encounters a *global suspension note* within a suspension note list of a local read-only variable (cf. Section 4.3.3).

Entries in the *Remote Instance Stack* are processed by the Distributor. Each entry contains the local address of a former variable, which has meanwhile become instantiated, together with the the global address of a related read-only variable or a related *XER* at a remote unit. By means of the local address, the Distributor identifies the resulting term to be copied to the remote unit.

Variable Migrations. When the Reducer encounters a XER in an attempt to unify two terms and the unification operation requires to read or write this XER, it activates the Distributor. The only necessary information for the Distributor to initiate a corresponding variable migration operation is the address of the local XER for which the variable is requested. This is obtained from the Reducer.

Local Synchronization. The described interaction between Reducer and Distributor requires that both subunits operate on common data structures. The Reducer as well as the Distributor, both must have access to the local *Resolvent* and to the *Remote Instance Stack.*

In order to avoid possible data conflicts between the concurrently operating subunits, mutual exclusive access to locally shared data structures must be guaranteed. Furthermore, it has to be ensured that both functional subunits cannot be interrupted by each other. That is, while performing a number of related operations, a unit must be able to maintain control over the shared data structures until a well defined state has been reached.

In particular, it is important that the Reducer cannot be interrupted during a *process reduction cycle.* The resulting variable bindings have either to be carried out all at once or none of it. In case that the reduction attempt results in a process suspension, the old state of the local resolvent has to be restored, i.e. any bindings effected by this reduction attempt must be undone immediately.

A switch of control from one unit to the other unit is realized by sending a *control signal* over the corresponding communication channel (see Figure 5.2). A control signal is always passed from the active unit to the inactive unit. While being inactive, a unit is not able to generate control signals but has to wait until the next signal has been received. On the other hand, as soon as the active unit has delivered control to the waiting unit, it enters a wait state.

Initially, the Distributor is active, while the Reducer is waiting to receive a *START* signal. When receiving the *START* signal, the Reducer starts its computation. Now, the Reducer maintains control until an events occurs which causes the Reducer to deliver control back to the Distributor by generating an according signal. The possible events together with the signals indicating them are listed below:

IDLE	:	The Reducer has run out of work. as all processes have been reduced, or all remaining processes have been suspended.
XER	:	A variable migration operation is required.
FAIL	:	A reduction failure has occurred.
SYNC	:	A well defined state has been reached, after a predefined number k of ($k \geq 1$) process reductions or process suspensions have been performed.

When receiving a control signal from the Reducer, the reaction of the Distributor depends on the particular signal type. Unless receiving a *FAIL* signal, the Distributor may want to restart the Reducer after the event has been handled and the computation has not yet been terminated. This is done by returning a *CONTINUE* signal back to the Reducer.

5.2.2 The Distributor Subunit

The Distributor acts as the local communication control unit. Beside running the desired network protocols for the Reducer, it initiates and controls all necessary activities concerning the configuration of the parallel machine network, dynamic load balancing, distributed termination detection, and distributed debugging.

In order to carry out the above protocols, the Distributor needs to handle various kinds of messages that are either sent by the host unit or any of the other reduction units. Messages addressing the local reduction unit are first received by the Router, which then directs them to the Distributor. Similarly, the Distributor delivers messages to be sent to remote units to the Router.

The interface between the Distributor and the Router is realized by means of two unidirectional communication channels. One is directed from the Router to the Distributor and the other one from the Distributor to the Router. Network messages that may be sent or received by the Distributor are classified according to the following nine message categories:

- o *Variable Request Messages (VRMs)*

- o *Variable Grant Messages (VGMs)*

- o *Variable Location Messages (VLMs)*

- o *Remote Instance Messages (RIMs)*

- o *Work Load Request Messages (LRMs)*

- o *Work Load Grant Messages (LGMs)*

- o *Global Termination Messages (GTMs)*

- o *Network Boot Messages (NBMs)*

- o *Debugging Messages (DMs)*

Variable Request, Variable Grant and Variable Location Messages. *VRMs* as well as *VGMs*, both result from the variable migration protocol. A variable migration operation is initiated by sending a *VRM* to the corresponding variable owner. If a *VRM* is received by a reduction unit no more being the current variable owner, i.e. the requested variable has already migrated to some other reduction unit, then the *VRM* is forwarded accordingly.

When receiving a *VRM* for an unbound variable, the Distributor responds with a *VGM*. At the same time, the local variable becomes replaced by a *XER* pointing to the new variable location. Otherwise, if the variable has already been instantiated, the Distributer generates a *RIM* , in order to copy the resulting term.

A *VLM* indicates the *remote locations* of one or more newly created variable instances resulting from a process migration operation. A *VLM* refers to write-enabled as well as read-only variables at the reduction unit that has received the process structures. The addresses contained in the *VLM* are used to attach the newly created instances to already existing distributed variable representations.

Remote Instance Messages. A *RIM* always contains a single term together with the address of a read-only variable or a *XER*. If it refers to a *XER*, the *RIM* has been sent in response to a *VRM*.

In case that the *RIM* refers to a local read-only variable, the corresponding write-enabled variable has been instantiated on a remote reduction unit. The *RIM* causes an instantiation of the local read-only variable, which also has the effect that the attached suspension note list is processed as described in Section 4.3.3.

Load Request and Load Grant Messages. Depending on the applied load balancing policy, a *LRM* may have one of two possible meanings. First, it may represent a request for more work from a reduction unit which is going to run out of work. Second, it may represent a status request about the current work load situation.

When receiving a *LGM*, it may contain any number r $(r \geq 1)$ of FCP processes to be included into to the local resolvent.

Global Termination Messages. There is just one type of global termination message. A *GTM* corresponds to a termination detection token, which has been generated by the distributed termination detection algorithm. The Distributor handles termination detection tokens as explained in Section 4.4.

Network Boot and Debugging Messages. There are various kinds of *NBMs* as well as *DMs*. *NBMs* deal with the configuration and initialization of the parallel machine network. *NBs* provide necessary information required for distributed debugging. As the meaning of these messages is not material to the understanding of the parallel machine, the details are skipped here.

The Distributor Cycle. Within its main execution cycle, as illustrated below, the Distributor first waits for the Reducer to pass the next control signal over the *fromReducer* channel. As soon as a signal has arrived, the Distributor is reactivated.

The most frequently generated control signals are the signals *SYNC* and *XER*. Both signals enable the Distributor to process a number of received network messages. In addition to the network messages itself, which are passed via the channel *fromRouter*, the Router also provides the total number of currently received network messages by the variable *message_count*. The current value of this variable when entering the network message loop thereby determines the number of messages that are processed at once. Subsequently received messages will be delayed until the next iteration step in the main execution cycle.

At the end of the Distributor cycle, control is returned to the Reducer by passing a *CONTINUE* signal over the *toReducer* channel.

```
channel *toReducer, *fromReducer, *toRouter, *fromRouter;

...

while ( not( GlobalTermination )) /* Main Distributor cycle */
{
     signal= *fromReducer;         /* Wait for control signal from Reducer */
     switch ( signal.signal_type )
     {
       case SYNC: ... break;
       case XER : ... break;       /* Generate a corresponding VRM        */
       case IDLE: ... break;       /* Enter termination detection cycle */
       case FAIL: ...              /* Abort computation, enter debugger */
     }

     if ( message_count > 0 )
     { /* Read network messages */

        for ( i=0; i< message_count; i++ )
        {
           message= fromRouter; /* Dequeue next message */
           switch ( message.message_type )
           {
             case VRM: ... break;
             case VGM: ... break;
             case RIM: ... break;
             case LRM: ... break;
             case LGM: ... break;
             case GTM: ... break;
             case NBM: ... break;
             case DM : ...
           }
        }
     }

     if ( remote_instance_count > 0 )
     { /* Process entries from RemoteInstanceStack */

        for ( i=0; i< remote_instance_count; i++ )
        {   /* Generate a RIM using data on top of RemoteInstanceStack */

           ...

           RemoteInstanceStack= pop( RemoteInstanceStack )
        }
     }

     ... /* Work load balancing activities */

     *toReducer= CONTINUE          /* Return control to Reducer */
}                                  /* End of Distributor cycle */
```

Handling of Variable Requests. A somewhat more detailed view on the handling of *VRMs* by the *network message loop* of the Distributor's main execution cycle is presented below. Thereby, it is assumed that the received *VRMs* have the following form: *message(VRM, LocalUnit, RemoteUnit, LocalAddress, RemoteAddress)*

```
while ( not( GlobalTermination )) /* Main Distributor cycle */
{

    ...

    message= *fromRouter; /* Dequeue next message */
    switch ( message.message_type )
    {
      case VRM:
            switch ( dataType( LocalAddress )) /* LocalAddress refers to
            {                                     the requested variable */
              case VAR:
                        /* The requested variable does locally exists. */
                        sendMessage(VGM, RemoteUnit, LocalUnit,
                                RemoteAddress, Value(LocalAddress) );

                        /* Value(LocalAddress) represents the pointer
                           to a possibly existing read-only variable. */

                        Type(LocalAddress)= XER;
                        Value(LocalAddress)= <RemoteUnit,RemoteAddress>;
                        break;

              case XER: /* As the requested variable is no more local,
                           the VRM becomes forwarded accordingly, i.e.
                           to the location identified by the local XER. */

                        sendMessage(VRM,Value(LocalAddress).UnitID,RemoteUnit,
                                Value(LocalAddress).AddressID,RemoteAddress);
                        break;

              default:  /* The requested variable has become instantiated. */
                        RemoteInstanceStack= push( RemoteInstanceStack,
                            (LocalAddress,<RemoteUnit,RemoteAddress>) );
            }   break;


      case VGM: ... break;
      case RIM: ... break;
      ...
      case DM : ...
    }

    ...

} /* End of Distributor cycle */
```

5.2.3   The Router Subunit

According to the distinction of communication depending on the functional layer it is performed, the Router subunit deals with physical or system layer communication. There arise two basic tasks which are associated with this subunit. First, the Router carries out and controls any global message transfer operation in which a reduction unit is involved. Second, it performs the adaptation of asynchronous communication at the logical communication layer to the underlying synchronous communication architecture, and vice versa.

As already mentioned, the Router is implemented by a set of parallel processes co-operatively providing the intended function. The internal structure of the Router, as illustrated in Figure 5.3, consists of 6 parallel processes. The corresponding submodules are represented by the *Routing Control Unit (RCU)*, the gates *Gate0,...,Gate3*, and the *InGate* to the local Distributor.

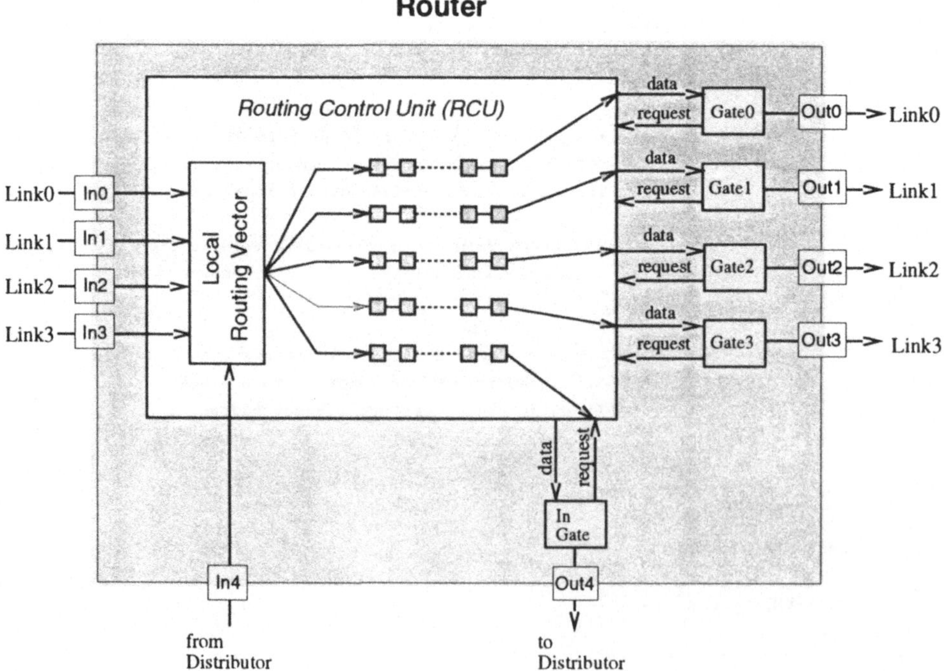

Figure 5.3: Router Architecture.

Receiving Messages. The RCU handles the stream of received messages. This includes messages arriving at one of the four external ports $In_0, ..., In_3$ as well as messages from the local Distributor. Depending on the address identifier contained in the message header, the RCU directs the messages either to the local Distributor or to one of the external ports $Out_0, ..., Out_3$.

Using the topology information which is encoded in the local routing vector, messages addressing other processing elements, i.e. any of the other reduction units or the host unit, are forwarded via a shortest path. For each existing unit identifier l $(0 \leq k \leq n)$, where the identifier 0 refers to the host unit, the local routing vector rv_k of a reduction unit RU_k $(1 \leq k \leq n)$ identifies a suitable message queue $rv_k[l] = i$ $(0 \leq i \leq 3)$ attached to the corresponding port Out_i.

A particular important issue concerning the implementation of the RCU is *fairness*. It must be avoided that a sequence of messages arriving on a particular port can pass immediately, while those messages arriving at any of the other ports have to wait for an indefinite period of time. For that reason, the RCU serves the external ports $In_0, ..., In_3$ together with the Distributor port In_4 in a round robin fashion. At most one message is received at a port before the RCU switches over to the next one. Nevertheless, it switches over to the next port, immediately, whenever a requested port is currently inactive. A fair handling is guaranteed, as the order in which the ports are requested by the RCU is permuted within each iteration cycle.

Sending Messages. From the internal message queues within the RCU the messages, one by one, are passed over to the corresponding gate processes. The gate processes, which have access to the Transputer links, then perform the actual send operations. Each of the gate processes $Gate0, ..., Gate3$ as well as the $InGate$ can handle one message at a time. As soon as a gate process has completed its current send operation, it requests the next message from the RCU by generating a control signal over the *request channel*. If the corresponding message queue still contains further messages, the RCU passes the next message over the *data channel*.

Similar to the above discussed fairness requirements concerning the receive operations, fairness is required for send operations as well. In particular, it must not happen that requests of a certain gate process are frequently served, while requests of other gate processes are ignored. In order to eliminate this problem, the RCU accepts requests from gate processes in a round robin fashion.

While the gate processes are controling the send operations, the RCU is able to handle newly received messages. This kind of functional separation of the Router subunit into a number of independently operating processes in combination with the applied message buffer queues yields the desired functional behaviour.

5.3   Host Unit Architecture

Similar to the described realization of reduction units, the host unit also divides
into a number of submodules which are embedded into parallel processes running
on the same Transputer node. However, the only functional component which the
host unit has in common with the reduction units is the Router subunit. Instead of
the Distributor and the Reducer subunits, the host unit is supplied with a number
of special functional components not required for reduction units. These additional
components, in particular, provide the necessary facilities for network booting, program loading, performing I/O operations, generating run-time statistics, distributed
debugging, controlling dynamic load balancing, and global termination detection (see
Figure 5.4).

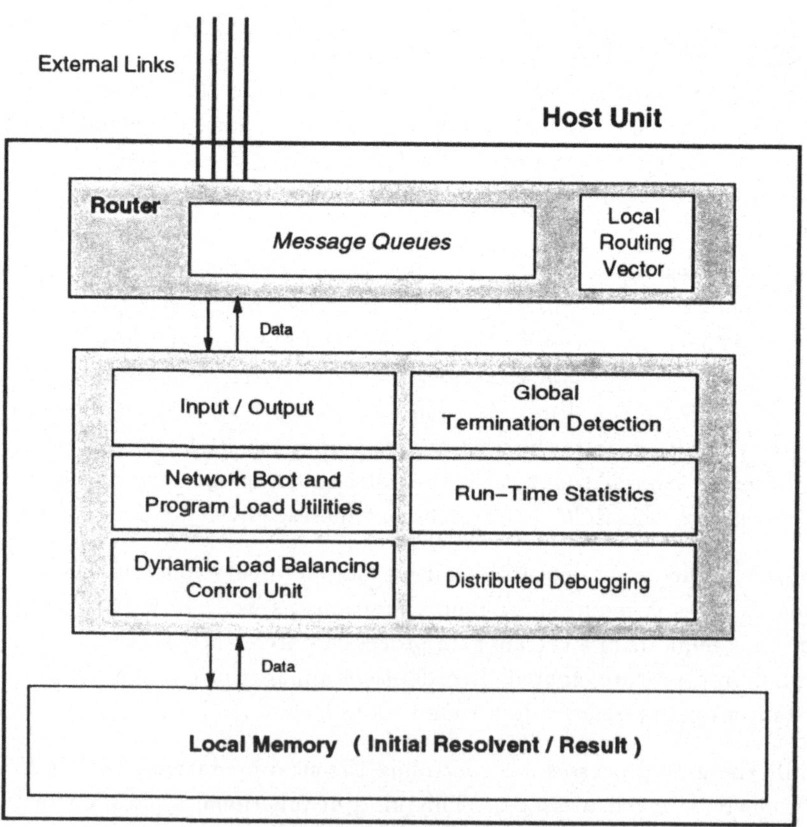

Figure 5.4: Host Unit Architecture

With respect to the described configuration scheme of the parallel FCP machine on a Transputer network, a configuration consisting of n reduction units always requires $n+1$ processors, since an extra processor is allocated to run the host unit. This configuration scheme was chosen for symmetry reasons, only. Alternatively, it would also be possible to integrate the host unit together with one of the reduction units on the same processor.

All such an integrated approach would require, is a slightly modified Router subunit with two additional internal ports and a correspondingly extended local routing vector. In fact, this might provide a reasonable solution, as the host unit is primarily busy during the network boot phase, while it is almost idle most of the time during a computation.

Chapter 6

Performance Measurements and Optimizations

The current prototype implementation of the parallel FCP machine comprises about 13.000 lines of Par.C code not including the evaluation tools. The performance measurements have been carried out on a *Parsytec Supercluster* using Par.C System Version 1.22 [Parsec89]. The hardware platform is a fully reconfigurable Transputer system architecture. The present system offers network configurations with up to 320 T800 Transputer nodes, each of which has direct access to 4 MB local RAM. A system of this kind consists of 20 clusters each containing 16 processors together with a 96×96 crossbar switch. In a second stage of the network, 8 switches connect 32 external edges of each cluster. A detailed description of the hardware system architecture is presented in [Funke92].

The compiler used to translate FCP application programs into executable machine programs has been generated using the *Eli* compiler construction system [Gray92] and runs on SUN workstations. For the compiler, it is not necessary to know the parallel machine architecture. In fact, the same machine program may be executed on any number of Transputer nodes as well as on a purely sequential version of the FCP machine running on SUN workstations. In its current version, the compiler performs only a few basic optimizations.

6.1 Performance Measures

For the two benchmark programs *Towers of Hanoi* [Houri87] and *Matrix-Multiplication* [Shapiro89], which are listed below, the resulting speedup behaviour is presented in Figure 6.1 and Figure 6.2. *Speedup* here is defined as the number of times faster than a single-processor system an n-processor system runs a given problem. More precisely, the term single-processor system thereby refers to an instance of the parallel FCP machine consisting of the host unit together with a single reduction unit. Accordingly, an n-processor system employs n reduction units in addition to the host unit. The diagrams show the above relationship not only for different network sizes but also various problem sizes. The underlying performance measures are mean values that rely on extensive performance measurements.

The network topology of all networks used for the program *Towers of Hanoi* respectively has been a de Bruijn-type network [Leighton92] of the appropriate dimension. The network topology of all networks used for the program *Matrix-Multiplication* respectively has been a torus of the appropriate dimension. The dynamic load balancing policy with *Towers of Hanoi* was the one described in [Lueling92], while *Matrix-Multiplication* applied a simple neighbourhood-model.

```
% Towers of Hanoi

% hanoi(N,From,To,Moves) <-
% Moving N disks from position 'From' to position 'To'
% requires a sequence of moves as specified by 'Moves'.
%
% The resulting sequence 'Moves' is composed of a sub-
% sequence 'Before' removing N-1 disks from position
% 'From', a single move (From,To) applied to disk N, and
% a subsequence 'After' placing N-1 disks back on disk N.


hanoi(N,From,To,(Before,(From,To),After)) <-
  N>1 |
  sub(N,1,N1),
  free(From,To,Free),
  hanoi(N1,From,Free?,Before),
  hanoi(N1,Free?,To,After).

hanoi(1,From,To,(From,To)).


free(a,b,c).
free(a,c,b).
free(b,a,c).
free(c,a,b).
free(b,c,a).
free(c,b,a).
```

```
% Matrix-Multiplication:

% mm(Xm,Ym,Zm) <-
% Zm is the result of multiplying the matrix Xm with
% the transposed matrix Ym.


mm([Xv|Xm],Ym,[Zv|Zm]) <-
   vm(Xv,Ym,Zv),
   mm(Xm,Ym,Zm).
mm([],_,[]).

vm(Xv,[Yv|Ym],[Z|Zv]) <-
   ip(Xv,Yv,0,Z),
   vm(Xv,Ym,Zv).
vm(_,[],[]).

ip([X|Xs],[Y|Ys],P,S) <-
   P1 := P+X*Y,
   ip(Xs,Ys,P1?,S).
ip([],[],P,P).
```

Although one could argue that the absolute speedup values are not really impressive, the important result is that the speedup behaviour scales well with the problem size. That means, if the network size as well as the problem size is increased, the speedup also increases.

However, the speedup behaviour of the program *Towers of Hanoi* is considerably better than that of the program *Matrix-Multiplication*. Essentially, there are two reasons effecting this behaviour. First, *Towers of Hanoi* spawns significantly more processes than *Matrix-Multiplication*. The relationship is given by $O(2^{n+1})$ process creations to $O(n^3)$ process creations, where n respectively denotes the number of disks to be moved and the dimension of the two $n \times n$-matrices to be multiplied. At the same time, the processes created by *Towers of Hanoi* are relatively small and of constant size, while the process size of *Matrix-Multiplication* linearly increases with the value of n.

Second, the sequential FCP machine components use a special optimization applicable to numeric calculations. Instead of spawning a new process for each atomic calculation, e.g. 'P1 := P+X*Y', the result is always calculated immediately, provided that the required input values are available.

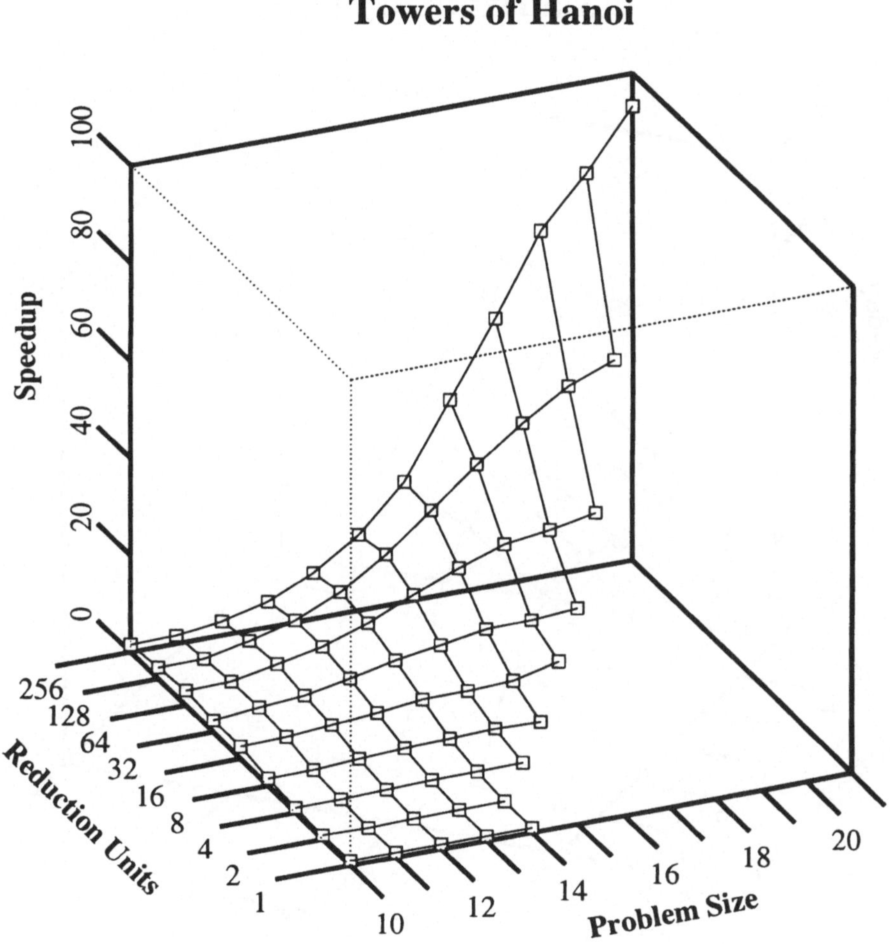

Figure 6.1: Speedups for the program *Towers of Hanoi*

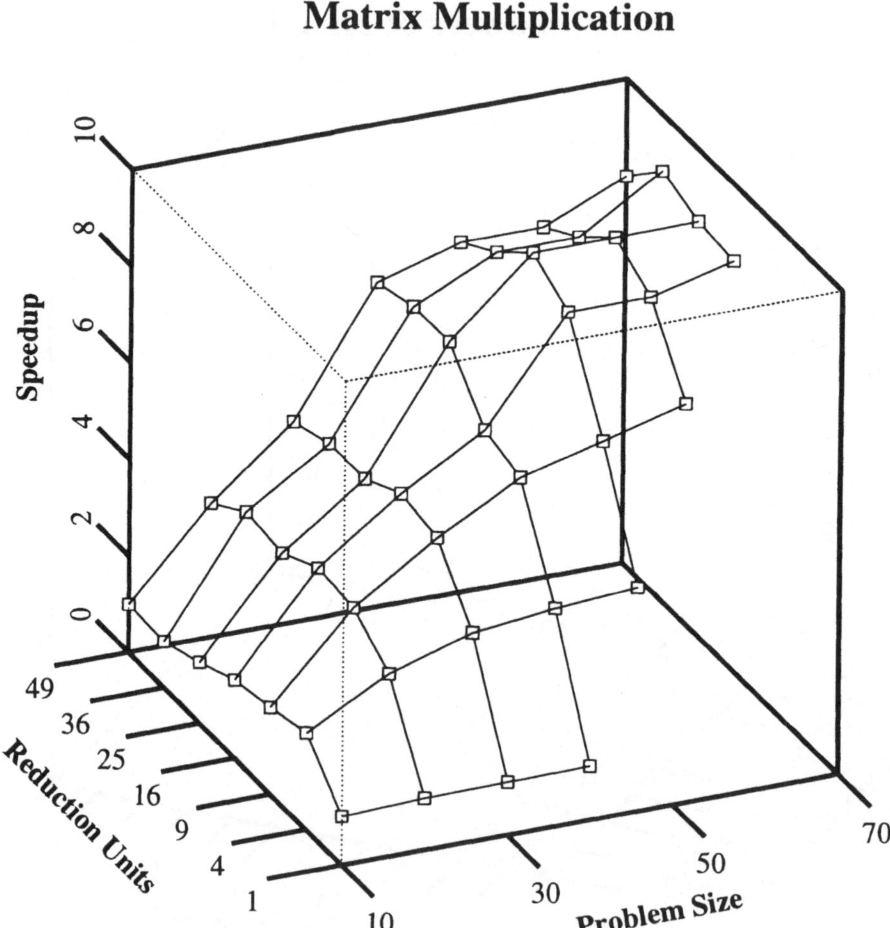

Figure 6.2: Speedups for the program *Matrix-Multiplication*

6.2 Possible Optimizations

Although the current prototype of the parallel FCP machine has been carefully designed and implemented, there remain a number of possible optimizations still to be done. First, of all these are optimizations concerning the compiler and the domain of dynamic load balancing. At the same time, certain compiler optimizations would also affect the organization of the sequential machine components.

Compared to the increase in performance that may be gained by compiler optimizations, the problem of dynamic load balancing, especially on large Transputer networks, seems to be even more important. With respect to both dimensions of dynamic load balancing, the load balancing policy as well as the process selection policy, more flexible algorithms dynamically adapting various control parameters to changing load situations are required.

With regard to the circumstances under which the impact of dynamic load balancing strategies on the system performance has been investigated so far, it should be realized that such toy programs as *Towers of Hanoi* running on a network with 256 processor nodes provide an extreme situation. There is much evidence, that the problem of dynamic load balancing looks much better when running the system with real programs in a multiprogramming mode.

Chapter 7

Conclusions

For the concurrent logic programming language FCP, a distributed implementation on large-scale Transputer systems has been described. The implementation is based on the concept of a *virtual parallel machine* executing compiled FCP code. Even though this parallel FCP machine was designed to run on a specific target architecture, which is a fully reconfigurable multi-Transputer system, the same implementation also runs on any other Transputer network provided that the underlying Par.C system is available. Moreover, most of the concepts presented for implementing the parallel FCP machine on Transputer networks are general in the sense that they should be applicable to the whole class of message-passing multiprocessor systems.

The parallel FCP machine. The primary objective in the design of the parallel FCP machine was to obtain an efficient embedding of the parallel execution model of FCP on Transputer networks. Additional requirements were to achieve maximum scalability and independence of the network topology.

The proposed parallel machine design involves two different layers of abstraction dealing with basically different kinds of problems. The more general concepts concerning issues like distributed data representation, global communication and synchronization, dynamic load balancing, etc. are discussed in terms of an abstract system architecture.

An embedding of the abstract system architecture onto a real hardware system then is specified in a subsequent design step. This is concerned with low-level communication and synchronization tasks, such as global message routing or synchronization features ensuring the atomicity of reduction operations.

The resulting functional architecture is oriented towards an efficient integration of the different communication models for the application language and the Transputer hardware. The final system, as it is realized by the implemented prototype, can be characterized by a number of basic design concepts:

o *Overall parallelization rests upon the MIMD scheme:* The parallel machine is realized through a network of asynchronously operating sequential FCP machines. By partitioning the global resolvent into a number of local subresolvents, the sequential machines cooperatively perform a distributed computation according to the MIMD scheme.

o *Physically distributed system with logically shared memory:* Although there is no shared global memory and the sequential FCP machine components may communicate by message-passing, only, they may interact with each other without being aware of it. In order to obtain a clear separation between tasks devoted to sequential program execution and tasks devoted to parallelization, the operations required for global interactions are completely transferred to the underlying run-time system.

o *Compiled FCP code is executed by abstract machines:* The sequential machine components are realized by abstract machines executing compiled FCP code. Except for a slightly extended unification algorithm, the applied machine components are almost identical to purely sequential FCP machines.

o *Effective parallelization appears transparent:* The above concepts together with the capability to perform dynamic load balancing, automatically, ensure an important feature: any effective parallelization appears transparent to the application programs.

o *Scalability and Adaptability:* The proposed algorithm for the parallel machine is generic in the sense that it is capable to generate suitable parallel machine instances dynamically adapting to any (connected) network topology for networks of any size, i.e. comprising any number n ($n \geq 2$) of processors.

Implementation on Transputers. Although the current prototype implementation has not been highly optimized, it already demonstrates the suitability of Transputer systems as target architectures for concurrent logic programming languages. An additional aspect encouraging this approach comes from the enhancements of Transputer communication facilities due to new technologies and further developments ([May90],[INMOS91]). Especially, such features as *hardware supported routing* facilities, support of *virtual links*, and a general increase in communication speed will match the communication demands of the proposed parallel machine design.

The programming language FCP. From the experience that have been made with the concurrent logic programming language FCP, it seems that dynamic dataflow synchronization based on read-only unification is an elegant solution but it comes for relatively high costs. Read-only test unification prior to commitment to a clause is often more than needed and less complex synchronization mechanisms (e.g. input matching as applied in Flat GHC), which can be implemented much more efficiently, would be sufficient for most applications.

On the other hand, the expressiveness a concurrent logic programming language should provide, strongly depends on the particular class of application problems to be implemented by this language. In order to investigate fundamental parallelization concepts, it might therefore be a better approach to select a more expressive language, which then can be further restricted according to specific application demands, than to add more and more artificial constructs to an implementation of a less expressive language.

Bibliography

[Bal88] BAL, H. E., AND TANENBAUM, A. S. 1988. Distributed programming with shared data. In *Proceedings of the IEEE International Conference on Computer Languages* (Miami), pp. 82-91.

[Bal89] BAL, H. E., STEINER, J. G., AND TANENBAUM, A. S. 1989. Programming languages for distributed computing systems. In *ACM Computing Surveys 21*, 3, pp. 261-322.

[Benker89] BENKER, H. ET AL. 1989. The knowledge crunching machine at ECRC: A joint R&D project of a high speed prolog system. In *ICL Technical Journal*, Nov. 1989, pp. 737-753.

[Bernstein81] BERNSTEIN, P. A., AND GOODMAN, N. 1981. Concurrency control in distributed database systems. In *Computing Surveys 13*, 2, pp 185-221.

[Bowen81] BOWEN, D. L., BYRD, L., PEREIRA, L. M., PEREIRA, F. C. N., AND WARREN, D. H. D. 1981. PROLOG on the DECSystem-10 user's manual. Technical Report, Dept. of Artificial Intelligence, University of Edinburgh, Scotland.

[Clark84] CLARK, K. L., AND GREGORY, S. 1984. PARLOG: Parallel programming in logic. Research Report DOC 84/4, Dept. of Computing, Imperial College of Science and Technology, London.

[Clocksin84] CLOCKSIN, W. F., AND MELLISH, C. S. 1984. *Programming in Prolog.* Springer-Verlag, Berlin.

[Coffman71] COFFMAN, E. G., ELPHICK, M. J., AND SHOSHANI, A. 1971. System deadlocks. In *Computing Surveys 3*, 2, pp. 67-78.

[Conery87] CONERY, J. S. 1987. *Parallel Execution of Logic Programs.* Kluwer Academic Publishers, Boston.

[Dijkstra75] DIJKSTRA, E. W. 1975. Guarded commands, nondeterminacy and formal derivation of programs. In *Communications of the ACM 18*, 8, pp. 453-457.

[Dijkstra83] DIJKSTRA, E. W., FEIJEN, W. H. J., AND VAN GASTEREN, A. J. M. 1983. Derivation of a termination detection algorithm for distributed computations., In *Information Processing Letters 16*, 5, pp. 217-219.

[vanEmden76] VAN EMDEN, M. H., AND KOWALSKI, R. A. 1976. The semantics of predicate logic as a programming language. In *Journal of the ACM 23*, 4, pp. 733-742.

[vanEmden82] VAN EMDEN, M. H., AND DE LUCENA FILHO, G. J. 1982. Predicate logic as a language for parallel programming. In *Logic Programming*, K. L. Clark, and S.-A. Tärnlund, Eds. Academic Press, London, pp. 189-198.

[Flynn66] FLYNN, M. J. 1966. Very high-speed computing systems. In *Proceedings of the IEEE 54*, 12, pp. 1901-1909.

[Foster87] FOSTER, I., AND TAYLOR, S. 1987. Flat Parlog: A basis for comparison. In *International Journal of Parallel Programming 16*, 2, pp. 87-125.

[Foster88] FOSTER, I. 1988. Parallel implementation of Parlog. In *Proceedings of the International Conference on Parallel Processing* (St. Charles), Ill., Vol. II, pp. 9-16.

[Foster89] FOSTER, I. 1989. A multicomputer garbage collector for a single assignment language. In *International Journal of Parallel Programming 18*, 3, pp. 181-203.

[Foster90] FOSTER, I., AND TAYLOR, S. 1990. *Strand, New Concepts in Parallel Programming*. Prentice-Hall, Englewood Cliffs, New Jersey.

[Fuchi86] FUCHI, K., AND FURUKAWA, K. 1987. The role of logic programming in the Fifth Generation Computer Project. In *New Generation Computing 5*, 1, pp. 3-28.

[Funke92] FUNKE, R. ET AL. 1992. An optimized reconfigurable architecture for Transputer networks. In *Proceedings of the 25th Hawaii International Conference on Computer Sciences* (Hawaii), pp. 237-245.

[Gray92] GRAY, R. W. ET AL. 1992. Eli: A complete compiler construction system. In *Communications of the ACM 35*, 2, pp. 121-131.

[Gregory87] GREGORY, S. 1987. *Parallel Logic Programming in PARLOG: The Language and Its Implementation*. Addison-Wesley Publishing Company, Wokingham, England.

[Glaesser90a] GLÄSSER, U., KÄRCHER, M., LEHRENFELD, G., AND VIETH, N. 1990A. Flat Concurrent Prolog on Transputers. In *Proceedings of the IFIP Working Conference on Decentralized Systems* (Lyon), C. Girault and M. Cosnard, Eds. North-Holland, Amsterdam, pp. 183-194.

[Glaesser90b] GLÄSSER, U., KÄRCHER, M., LEHRENFELD, G., AND VIETH, N. 1990B. Flat Concurrent Prolog on Transputers. In *Journal of Microcomputer Applications: Special Issue on Transputer Applications 13, 1, pp. 3-18* (extended version of [Glaesser90a]).

[Glaesser90c] GLÄSSER, U., AND LEHRENFELD, G. 1990C. A distributed implementation of Flat Concurrent Prolog on Transputer architectures. In *Proceedings of the UNESCO Conference on Parallel Computing in Engineering and Engineering Education* (Paris), pp. 181-185.

[Glaesser91a] GLÄSSER, U., KÄRCHER, M., AND LEHRENFELD, G. 1991. Dynamische Partitionierung asynchroner Prozeßnetzwerke am Beispiel Paralleler Logischer Programmierung. To appear in *Proceedings of the TAT '91* (Aachen, FRG, Sep. 17-18, 1991), Informatik-Fachberichte, Springer-Verlag, Berlin.

[Glaesser91b] GLÄSSER, U., HANNESEN, G., KÄRCHER, M., AND LEHRENFELD, G. 1991. A distributed implementation of Flat Concurrent Prolog on multi-Transputer environments. To appear in *Proceedings of the First International Conference of the Austrian Center for Parallel Computation* (Salzburg, Sep. 29 - Oct. 02, 1991), Lecture Notes in Computer Science, Springer-Verlag, Berlin.

[Glaesser92] GLÄSSER, U. 1992. A distributed implementation of Flat Concurrent Prolog on multi-Transputer environments. In *Distributed Prolog*, P. Kacsuk and M. Wise, Eds. John Wiley & Sons Ltd., Chichester, pp. 287-309.

[van de Goor89] VAN DE GOOR, A. J. 1989. *Computer Architecture and Design.* Addison-Wesely, New York.

[Grunzig92] GRUNZIG, P. 1992. Konzepte zur Hardware-Realisierung der Reduktionseinheit einer parallelen FCP-Maschine (Diplomarbeit). Paderborn University, Dept. of Mathematics & Computer Science, Paderborn, FRG, Feb. 1992.

[Hannesen91] HANNESEN, G. 1991. Implementierung und Optimierung von sequentiellen und parallelen FCP-Maschinen (Diplomarbeit). Paderborn University, Dept. of Mathematics & Computer Science, Paderborn, FRG, June 1991.

[Harel85] HAREL, D., AND PNUELI, A. 1985. On the development of reactive systems. In *Logics and Models of Concurrent Systems*. K. R. Apt, Ed. Lecture Notes in Computer Science, Springer-Verlag, Berlin.

[Hoare78] HOARE, C. A. R. 1978. Communicating sequential processes. In *Communications of the ACM 21*, 8, pp. 666-677.

[Houri87] HOURI, A., AND SHAPIRO, E. 1987. A sequential abstract machine for Flat Concurrent Prolog. In *Concurrent Prolog: Collected Papers*. Vol. 2, E. Shapiro, Ed. MIT Press, Cambridge, Mass., pp. 513-574.

[Ichiyoshi87] ICHIYOSHI, N., MIYAZAKI, T., AND TAKI, K. 1987. A distributed implementation of Flat GHC on the Multi-PSI. In *Logic Programming – Proceedings of the Fourth International Conference on Logic Programming* (Melbourne), J.-L. Lassez, Ed. MIT Press Cambridge, pp. 257-275.

[INMOS91] INMOS LTD. 1991. *The T9000 Transputer Products Overview Manual.* INMOS Databook series, Marlow, UK.

[Kaercher92] KÄRCHER, M. 1992. Automatische Parallelisierung am Beispiel eines verteilten FCP-Interpreters (Diplomarbeit). Paderborn University, Dept. of Mathematics & Computer Science, Paderborn, FRG (available before July, 1992).

[Kleine Buening86] KLEINE BÜNING, H., AND SCHMITGEN, S. 1986. *PROLOG.* B. G. Teubner Stuttgart.

[Kliger88] KLIGER, S., YARDENI, E., KAHN, K., AND SHAPIRO, E. 1988. The language FCP(:,?). In *Proceedings of the International Conference on Fifth Generation Computer Systems* (Tokyo), Ohmsha Ltd. Tokyo, Springer-Verlag, Berlin, pp. 763-773.

[Kowalski79a] KOWALSKI, R. 1979A. Algorithm = logic + control. In *Communications of the ACM 22, 7, pp. 424-436.*

[Kowalski79b] KOWALSKI, R. 1979B. *Logic for Problem Solving.* North-Holland, Amsterdam.

[Kurfess91] KURFESS, F. 1991. *Parallelism in Logic: Its Potential for Performance and Program Development.* Verlag Vieweg, Braunschweig, FRG.

[Lehrenfeld90] LEHRENFELD, G. 1990. Konzeption und Implementierung einer parallelen FCP-Maschine (Diplomarbeit). Paderborn University, Dept. of Mathematics & Computer Science, Paderborn, FRG, Dec. 1990.

[Leighton92] LEIGHTON, F. T. 1992. *Introduction to Parallel Algorithms and Architectures: Arrays ∘ Trees ∘ Hypercubes.* Morgan Kaufmann Publishers, San Mateo, Calif.

[Lueling91] LÜLING, R., MONIEN, B., AND RAMME, F. 1991. Load balancing in large networks: A comparative study. In *Proceedings of the 3rd IEEE Symposium on Parallel and Distributed Processing* (Dallas), pp. 686-689.

[Lueling92] LÜLING, R., AND MONIEN, B. 1992. Load balancing for distributed branch & bound algorithms. In *Proceedings of the 6th International Parallel Processing Symposium* (Beverly Hills), pp. 543-549.

[May85] MAY, D., AND SHEPHERD, R. 1985. Occam and the Transputer. In *Concurrent Languages in distributed Systems*. G. L. Reijens, and E. L. Dagless, Eds. North-Holland, Amsterdam, pp. 19-33.

[May90] MAY, D. 1990 Future directions in Transputer technology. In *Proceedings of UNESCO Conference on Parallel Computing in Engineering and Engineering Education* (Paris), pp. 193-203.

[Mierowsky85] MIEROWSKY, C., TAYLOR, S., SHAPIRO, E., LEVY, J., AND SAFRA, S. 1985. The design an implementation of Flat Concurrent Prolog. Technical Report CS85-9, Dept. of Computer Science, The Weizmann Institute of Science, Rehovot, Israel.

[Nakajima92] NAKAJIMA, K. 1992. Distributed implementation of KL1 on the Multi-PSI. In *Distributed Prolog*, P. Kacsuk and M. Wise, Eds. John Wiley & Sons Ltd., Chichester (To appear in June 1992).

[Okumura87] OKUMURA, A., AND MATSUMOTO, Y. 1987. Parallel programming with layered streams. In *Proceedings of the IEEE Symposium on Logic Programming* (San Francisco). IEEE New York, pp. 224-231.

[Parsec89] PARSEC 1989. *Par.C System: User's Manual and Library Reference Version 1.22*. Parsec Developments, Leiden, The Netherlands.

[Pnueli86] PNUELI, A. 1986. Applications of temporal logic to the specification and verification of reactive systems: A survey of current trends. In *Current Trends in Concurrency, Overviews and Tutorials*, Lecture Notes in Computer Science, Vol. 224, J. W. de Bakker, W.-P. de Roever, and G. Rozenberg, Eds. Springer-Verlag, New York, pp. 510-584.

[Ramme90] RAMME, F. 1990. Lastausgleichsverfahren in verteilten Systemen (Diplomarbeit). Paderborn University, Dept. of Mathematics & Computer Science, Paderborn, FRG, March 1990.

[Rokusawa88] ROKUSAWA, K., ICHIYOSHI, N., CHIKAYAMA, T., AND NAKASHIMA, H. 1988. An efficient termination detection and abortion algorithm for distributed processing systems. In *Proceedings of the International Conference on Parallel Processing*, Vol. I, pp. 18-22.

[Shapiro83] SHAPIRO, E. 1983. A subset of concurrent prolog and its interpreter. ICOT Technical Report TR-003, Institute for New Generation Computer Technology, Tokyo.

[Shapiro86] SHAPIRO, E. 1986. Concurrent Prolog: A progress report. *IEEE Computer 19, 8, pp. 44-58.*

[Shapiro89] SHAPIRO, E. 1989. The family of concurrent logic programming languages. In *ACM Computing Surveys 21*, 3, pp. 413-510.

[Silverman87] SILVERMAN, W., HIRSCH, M., HOURI, A., AND SHAPIRO, E. 1987. The Logix system user manual version 1.21. In *Concurrent Prolog : Collected Papers.* Vol. 2, E. Shapiro, Ed. MIT Press, Cambridge, Mass., pp. 46-77.

[Takeda90] TAKEDA, Y., NAKASHIMA, H., MASUDA, K., CHIKAYAMA, T., AND TAKI, K. 1990. A load balancing mechanism for large scale multiprocessor systems and its implementation. In *New Generation Computing*, 7, pp. 179-195.

[Takeuchi87] TAKEUCHI, A., AND FURUKAWA, K. 1987. Parallel logic programming Languages. In *Proceedings of the 3rd International Conference on Logic Programming* (London), Lecture Notes in Computer Science, Vol. 225, Springer-Verlag, New York, pp. 242-254

[Tanenbaum87] TANENBAUM, A. S. 1987 *Operating Systems: Design and Implementation.* Prentice-Hall, Englewood Cliffs, New Jersey.

[Taylor87] TAYLOR, S., SAFRA, S., AND SHAPIRO, E. 1987. A parallel implementation of Flat Concurrent Prolog. In *International Journal of Parallel Programming 15*, 3, pp. 245-275.

[Taylor89] TAYLOR, S. 1989. *Parallel Logic Programming Techniques.* Prentice-Hall, Englewood Cliffs, New Jersey.

[Treleaven82] TRELEAVEN, P. C., BROWNBRIDGE, D. R., AND RICHARD, P. H. 1982. Data-driven and demand-driven computer architecture. In *ACM Computing Surveys 14*, 1, pp. 93-143.

[Ueda86] UEDA , K. 1986. Guarded Horn clauses. In *Logic Programming.* Lecture Notes in Computer Science, Vol. 221, Springer-Verlag, Berlin, pp. 168-179.

[Ueda89] UEDA, K. 1989. Parallelism in logic programming. In *Proceedings of the IFIP Congress*, North-Holland, Amsterdam, pp. 957-964.

[Van Roy92] VAN ROY, P., AND DESPAIN, A. M. 1992. High-performance logic programming with the Aquarius Prolog Compiler. In *IEEE Computer 25*, 1, pp. 54-68.

[Warren83] WARREN, D. H. D. 1983. An abstract Prolog instruction set. Technical Note 309, Artificial Intelligence Center, SRI.

[Weinbaum87] WEINBAUM, D., AND SHAPIRO, E. 1987. Hardware description and simulation using Concurrent Prolog. In *Proceedings of the CHDL '87*, Elsevier Science Publishing, pp. 9-27.

[Xu90] XU, J., AND HWANG, K. 1990. Dynamic load balancing for parallel program execution on a message-passing multicomputer. In *Proceedings of the Second IEEE Symposium on Parallel and Distributed Processing* (Dallas), pp. 402-406.

Index